M336
Mathematics and Computing: a third-level course

GROUPS & GEOMETRY

UNIT GE3
TWO-DIMENSIONAL LATTICES

Prepared for the course team by
David Asche & Fred Holroyd

The Open University

This text forms part of an Open University third-level course.
The main printed materials for this course are as follows.

Block 1
Unit IB1   Tilings
Unit IB2   Groups: properties and examples
Unit IB3   Frieze patterns
Unit IB4   Groups: axioms and their consequences

Block 2
Unit GR1   Properties of the integers
Unit GR2   Abelian and cyclic groups
Unit GE1   Counting with groups
Unit GE2   Periodic and transitive tilings

Block 3
Unit GR3   Decomposition of Abelian groups
Unit GR4   Finite groups 1
Unit GE3   Two-dimensional lattices
Unit GE4   Wallpaper patterns

Block 4
Unit GR5   Sylow's theorems
Unit GR6   Finite groups 2
Unit GE5   Groups and solids in three dimensions
Unit GE6   Three-dimensional lattices and polyhedra

The course was produced by the following team:

Andrew Adamyk (BBC Producer)
David Asche (Author, Software and Video)
Jenny Chalmers (Publishing Editor)
Bob Coates (Author)
Sarah Crompton (Graphic Designer)
David Crowe (Author and Video)
Margaret Crowe (Course Manager)
Alison George (Graphic Artist)
Derek Goldrei (Groups Exercises and Assessment)
Fred Holroyd (Chair, Author, Video and Academic Editor)
Jack Koumi (BBC Producer)
Tim Lister (Geometry Exercises and Assessment)
Roger Lowry (Publishing Editor)
Bob Margolis (Author)
Roy Nelson (Author and Video)
Joe Rooney (Author and Video)
Peter Strain-Clark (Author and Video)
Pip Surgey (BBC Producer)

With valuable assistance from:

Maths Faculty Course Materials Production Unit
Christine Bestavachvili (Video Presenter)
Ian Brodie (Reader)
Andrew Brown (Reader)
Judith Daniels (Video Presenter)
Kathleen Gilmartin (Video Presenter)
Liz Scott (Reader)
Heidi Wilson (Reader)
Robin Wilson (Reader)

The external assessor was:

Norman Biggs (Professor of Mathematics, LSE)

The Open University, Walton Hall, Milton Keynes, MK7 6AA.

First published 1994. Reprinted 1997, 2002, 2007.

Copyright © 1994 The Open University

All rights reserved. No part of this publication may be reproduced, stored in a retrieval system or transmitted in any form or by any means, without written permission from the publisher or a licence from the Copyright Licensing Agency Limited. Details of such licences (for reprographic reproduction) may be obtained from the Copyright Licensing Agency Ltd of 90 Tottenham Court Road, London, W1P 9HE.

Edited, designed and typeset by the Open University using the Open University TEX System.

Printed in Malta by Gutenberg Press Limited.

ISBN 0 7492 2171 2

This text forms part of an Open University Third Level Course. If you would like a copy of *Studying with the Open University*, please write to the Central Enquiry Service, PO Box 200, The Open University, Walton Hall, Milton Keynes, MK7 6YZ. If you have not already enrolled on the Course and would like to buy this or other Open University material, please write to Open University Educational Enterprises Ltd, 12 Cofferidge Close, Stony Stratford, Milton Keynes, MK11 1BY, United Kingdom.

CONTENTS

Study guide		4
Introduction		5
1	**Lattices**	**6**
	1.1 Basic ideas	6
	1.2 The minimality conditions in the plane	13
2	**Symmetries of a plane lattice**	**18**
	2.1 Types of symmetry	18
	2.2 Symmetries which fix O	21
	2.3 Composites of symmetries	22
	2.4 The crystallographic restriction	25
3	**Five types of plane lattices**	**27**
	3.1 The parallelogram lattice	27
	3.2 The rectangular lattice	28
	3.3 The rhombic lattice	30
	3.4 The square lattice	31
	3.5 The hexagonal lattice	33
4	**The plane lattice groups**	**37**
	4.1 The parallelogram lattice	37
	4.2 The rectangular lattice	40
	4.3 The rhombic lattice	42
	4.4 The square lattice	43
	4.5 The hexagonal lattice	44
5	**The classification of plane lattices**	**45**
	5.1 The minimality conditions revisited	45
	5.2 Rectangular and rhombic indirect symmetries	46
	5.3 The classification theorem	50
Solutions to the exercises		53
Objectives		62
Index		63

STUDY GUIDE

Units GE3 and *GE4* follow on naturally from *Unit IB3*. Many of the concepts introduced in that unit will be needed here. We assume that you are familiar with affine transformations in the plane and know how the isometries of the plane are classified into geometric types. These topics are covered in *Unit IB1*, Sections 3 and 5. Other material which occurs in earlier units will be needed too. You should know the basic facts about groups and how groups can be specified. We shall consider various group actions, and you should understand the words *orbit* and *stabilizer* when they occur. These are discussed in *Unit IB2*, Section 5, and also in *Unit GE1*.

The five sections of this unit are roughly equal in length.

You will need the Isometry Toolkit card for your study of this unit, particularly for Sections 2 to 4. You will not need any of the other cards or overlays in the *Geometry Envelope*.

There is a video programme, VC3A *Lattices and Wallpaper Patterns*, associated with this unit and with *Unit GE4 Wallpaper Patterns*. If you have time, you may wish to view it before you start your study of *Unit GE3*. However, you will probably find it useful to view it again during your study of *Unit GE4*, so if you are short of time it would be advisable to postpone your viewing until then.

There is no audio programme associated with this unit.

INTRODUCTION

This unit is concerned with the concept of a two-dimensional *lattice*. We begin by stating what is meant by this geometrical object and then proceed to investigate the various symmetries that such a lattice may possess. After this, we show that there are five different types of two-dimensional (or *plane*) lattices, and we study the various groups of symmetries that may arise. Although lattices are interesting in themselves, the main purpose of our study is to prepare the groundwork for a classification of wallpaper patterns, which is the subject matter of *Unit GE4*. Both units are concerned with certain kinds of patterns drawn in the plane, and we shall be discussing how these patterns may be classified into geometric types.

You will recall from *Unit IB3* that frieze patterns are also drawn in the plane and we saw how these were classified into geometric types. The difference between the patterns that we study here and those in *Unit IB3* is that, while friezes admit translations in only one direction, our patterns admit translations in two independent directions.

Remember how we constructed frieze patterns. We started with a basic motif drawn in a rectangle (the base rectangle), and then applied horizontal translations whose magnitudes were multiples of the length of the rectangle. This gave us an infinite horizontal strip consisting of a design which repeats at regular intervals. The width of the strip is of no consequence. It could even extend so that the pattern occupied the whole plane, provided that the only translational symmetries are horizontal. The figure shown below is an example of a frieze pattern. It is, of course, supposed to extend indefinitely to the right and to the left.

By *length* of the rectangle we mean the length of a horizontal side; it is possible, of course, for the horizontal sides nevertheless to be shorter than the vertical sides.

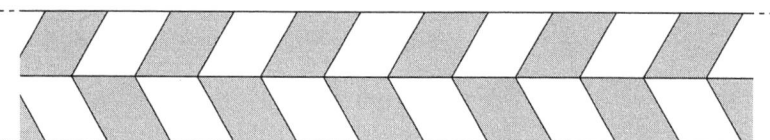

Figure 0.1

We shall find that wallpaper patterns can be constructed in a similar manner. Starting with a basic motif drawn in a parallelogram, we apply translations in directions parallel to the sides of the parallelogram and by magnitudes which are multiples of the lengths of the sides. Since these translations are performed in two independent directions, the pattern must occupy the whole plane and not merely a strip.

When we studied frieze patterns, we found that many of them possess symmetries in addition to the horizontal translations. Some patterns admit rotations, reflections or glide reflections as well. In the example shown in Figure 0.1, there are glide reflections but no rotations or reflections. In a similar way, many of our wallpaper patterns will have symmetries in addition to the translations. Again these will be rotations, reflections or glide reflections, but we shall find a greater variety of possibilities. In the case of frieze patterns, a study of the symmetries allowed us to classify them into seven different geometric types. By the end of our study of wallpaper patterns, we shall have found that there are seventeen types of wallpaper pattern distinguished by the kinds of symmetry which they possess.

In Figure 0.2, we see an example of a wallpaper pattern which has a lot of symmetry. There are rotations of different orders, there are reflections in many lines and there are glide reflections as well.

Figure 0.2

This unit is divided into five sections. In Section 1, we introduce the idea of a lattice (in particular, a two-dimensional or plane lattice), and show how it can be represented algebraically in terms of what is called a *basis*. In Section 2, we examine the various kinds of symmetries that a plane lattice may possess. In Section 3, we find five distinct geometric types of plane lattice. In Section 4, we study the symmetry groups of each of the five types of plane lattice and, finally, in Section 5, we prove that the five geometric types found in Section 3 are the only ones that occur.

1 LATTICES

1.1 Basic ideas

In *Unit IB1*, Section 3, we reminded you of the familiar algebraic model of the plane using a coordinate system. A vector $\mathbf{a} = (a_1, a_2)$ in this system has *length*

$$\|\mathbf{a}\| = \sqrt{\mathbf{a} \cdot \mathbf{a}}.$$

This length is equal to the distance from the origin O to the point A with coordinates a_1 and a_2. The zero vector $(0, 0)$ is denoted by $\mathbf{0}$.

In order to define and construct a two-dimensional lattice, we must choose two vectors $\mathbf{a}$ and $\mathbf{b}$ in $\mathbb{R}^2$ which are linearly independent. This means that every vector in $\mathbb{R}^2$ can be written in the form $x\mathbf{a} + y\mathbf{b}$ in exactly one way. It follows that neither of the vectors $\mathbf{a}$ and $\mathbf{b}$ is equal to $\mathbf{0}$, and that neither is a multiple of the other.

Although in this unit we sometimes consider linear combinations $x\mathbf{a} + y\mathbf{b}$ where x and y are not integers, we are mainly interested in what are called **integer combinations** of $\mathbf{a}$ and $\mathbf{b}$. These are vectors of the form $n\mathbf{a} + m\mathbf{b}$, where n and m are *required to be integers*.

This is a version for vectors of the concept of integer combination which you met in *Unit GR1*.

We can now define a *two-dimensional lattice* (more conveniently known as a *plane lattice*).

> *Definition 1.1 Two-dimensional (or plane) lattice*
>
> A **two-dimensional** (or **plane**) **lattice** is a set consisting of all points in the plane whose corresponding vectors constitute the set
>
> $$L(\mathbf{a}, \mathbf{b}) = \{n\mathbf{a} + m\mathbf{b} : n, m \in \mathbb{Z}\},$$
>
> where $\mathbf{a}$ and $\mathbf{b}$ are linearly independent vectors of $\mathbb{R}^2$.

Since all the lattices in this unit are two-dimensional, we shall frequently drop the term *two-dimensional* or *plane*. You should remember, though, that the concept of a lattice generalizes to n-dimensional space.

It is a bit tiresome to keep referring to '...points in the plane whose corresponding vectors are ...'. Once we have set up our coordinate system, we can identify points in the plane with the vectors which represent them. When we are talking geometrically, we speak of a lattice L and name its points using capital letters such as O, A and B. But when the discussion is algebraic we tend to talk about a lattice $L(\mathbf{a}, \mathbf{b})$ and use vectors such as $\mathbf{0}$, $\mathbf{a}$ and $\mathbf{b}$ to specify its points.

The word lattice is used differently in some other branches of mathematics so, if you come across it in books, do not expect it to have the same meaning that we give it here. Some authors call our lattice a *point lattice* or *vector lattice*.

Exercise 1.1

Suppose that $\mathbf{p}$ and $\mathbf{q}$ are two points of the lattice $L(\mathbf{a}, \mathbf{b})$. Show that any integer combination $n\mathbf{p} + m\mathbf{q}$ is also a point of the lattice $L(\mathbf{a}, \mathbf{b})$.

The above definition of a lattice is given algebraically. In order to view it geometrically, it is useful to consider the translations of the plane which map the origin O to any other point of the lattice. Starting with a lattice $L = L(\mathbf{a}, \mathbf{b})$, we shall consider the translations $t[\mathbf{a}]$ and $t[\mathbf{b}]$ of the plane which correspond to these vectors $\mathbf{a}$ and $\mathbf{b}$. If $\mathbf{a}$ represents the point A and $\mathbf{b}$ represents the point B, then the translation $t[\mathbf{a}]$ will map O to A and the translation $t[\mathbf{b}]$ will map O to B (see Figure 1.1). We can express this by writing $\mathbf{a} = t[\mathbf{a}](\mathbf{0})$, and $\mathbf{b} = t[\mathbf{b}](\mathbf{0})$. If we first apply $t[\mathbf{a}]$ to $\mathbf{0}$ and then apply $t[\mathbf{b}]$ to $t[\mathbf{a}](\mathbf{0})$, we arrive at the point represented by the vector $\mathbf{a} + \mathbf{b}$. Since $\mathbf{b} + \mathbf{a} = \mathbf{a} + \mathbf{b}$, we have $t[\mathbf{a}]\,t[\mathbf{b}] = t[\mathbf{b}]\,t[\mathbf{a}]$ and, more generally, all possible composites of $t[\mathbf{a}]$ and $t[\mathbf{b}]$ can be written in the form $(t[\mathbf{a}])^n\,(t[\mathbf{b}])^m$. Such a translation will map the point O to the point in the plane represented by the vector $n\mathbf{a} + m\mathbf{b}$. If we apply all these translations to the point O, we obtain the lattice L.

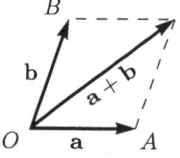

Figure 1.1

In Figure 1.2 below, we show what a typical lattice looks like.

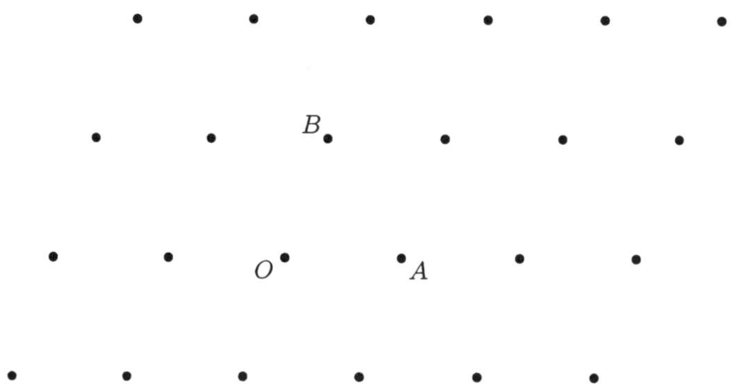

Figure 1.2

We have displayed our origin O and two points A and B which lie on different lines through O. The points on the line through O and A are obtained by applying to O the various integer powers of the translation which maps O to A. In a similar way, we obtain all the points on the line through O and B. The remaining points of the lattice are obtained by forming composites of these translations.

We have, of course, shown only part of the lattice. You should imagine that it continues indefinitely in all directions. A lattice can be thought of as the set of crossing points of two sets of equally spaced parallel lines such as you might find in some lattice work or a garden trellis.

If we are given a lattice such as the one above, with a particular point selected for the origin, we can find an algebraic expression for it by selecting two suitable points A and B and forming the set $L(\mathbf{a}, \mathbf{b})$, where $\mathbf{a}$ represents A and $\mathbf{b}$ represents B. Now, clearly, not *every* pair of points A and B will be a suitable choice. On the other hand, there are many pairs of points which *would* be suitable. We shall look at an example next.

Example 1.1

Take the lattice whose points belong to the set $\{(n,m): n,m \in \mathbb{Z}\}$. This lattice is called the **lattice of integer points** in the plane. The x-coordinate and the y-coordinate of each point are both integers. We must now find two suitable vectors $\mathbf{a}$ and $\mathbf{b}$ so that we can describe the lattice algebraically. The most obvious choice is to take $\mathbf{a} = (1,0)$ and $\mathbf{b} = (0,1)$. Then a general point (n,m) of the lattice can be expressed as $n\mathbf{a} + m\mathbf{b}$. The lattice can therefore be specified algebraically as $L(\mathbf{a}, \mathbf{b})$. But these choices for $\mathbf{a}$ and $\mathbf{b}$ are by no means unique. Suppose that we took $\mathbf{b} = (1,1)$ instead of $(0,1)$. We find that a general point (n,m) of the lattice can be expressed in terms of the new choice of $\mathbf{a}$ and $\mathbf{b}$, since we have

$$(n,m) = (n-m)(1,0) + m(1,1),$$

as you may verify. ◆

The following exercise concerns the lattice of integer points that we have been discussing. You may find it helpful to make a sketch of the lattice.

Exercise 1.2

Which of the following are suitable choices for $\mathbf{a}$ and $\mathbf{b}$?

(a) $\mathbf{a} = (1,0)$, $\mathbf{b} = (2,1)$
(b) $\mathbf{a} = (1,0)$, $\mathbf{b} = (1,2)$
(c) $\mathbf{a} = (0,1)$, $\mathbf{b} = (-1,3)$
(d) $\mathbf{a} = (2,3)$, $\mathbf{b} = (1,1)$

It will be useful to have a name to describe a pair of vectors from which we can obtain all the points of a lattice by forming linear combinations with integer coefficients. Such a pair of vectors is called a *basis* for the lattice.

Definition 1.2 Basis for a (plane) lattice

A pair of vectors $\{\mathbf{a}, \mathbf{b}\}$ is said to be a **basis** for a (plane) lattice L if the vectors $\mathbf{a}$ and $\mathbf{b}$ are linearly independent and

$$L = L(\mathbf{a}, \mathbf{b}) = \{n\mathbf{a} + m\mathbf{b}: n,m \in \mathbb{Z}\}.$$

If a pair $\{\mathbf{a}, \mathbf{b}\}$ of vectors forms a basis for a lattice L, the definition states that $\mathbf{a}$ and $\mathbf{b}$ are linearly independent. This means that every point in $\mathbb{R}^2$ has a unique expression of the form $x\mathbf{a} + y\mathbf{b}$, where x and y are *real* numbers. If you have met vector spaces in your previous studies, you will see that $\{\mathbf{a}, \mathbf{b}\}$ is a basis for $\mathbb{R}^2$ in the usual sense. Lattices, however, are not vector spaces since their points are only *integer* combinations of $\mathbf{a}$ and $\mathbf{b}$.

For the lattice shown in Figure 1.2, we took one of the lattice points to be our origin O and selected two points A and B whose corresponding vectors $\mathbf{a}$ and $\mathbf{b}$ gave us a basis $\{\mathbf{a}, \mathbf{b}\}$ for the lattice. Given a lattice, we may choose any lattice point to be our origin O; but how does one choose suitable points A and B?

To answer this question, we first observe that the triangle OAB contains no lattice points other than O, A and B. The following theorem shows that this condition is both necessary and sufficient for the pair $\{\mathbf{a}, \mathbf{b}\}$ to be a basis for the lattice.

> **Theorem 1.1**
>
> A linearly independent pair of vectors $\{\mathbf{a}, \mathbf{b}\}$ in a plane lattice L is a basis for L if and only if the corresponding triangle OAB contains no lattice points other than O, A and B.

Proof

Let us first assume that $\{\mathbf{a}, \mathbf{b}\}$ is a basis for the lattice L (see Figure 1.3). The points of $\mathbb{R}^2$ which are contained in the triangle OAB are given by $x\mathbf{a} + y\mathbf{b}$, where $0 \leq x$, $0 \leq y$ and $x + y \leq 1$.

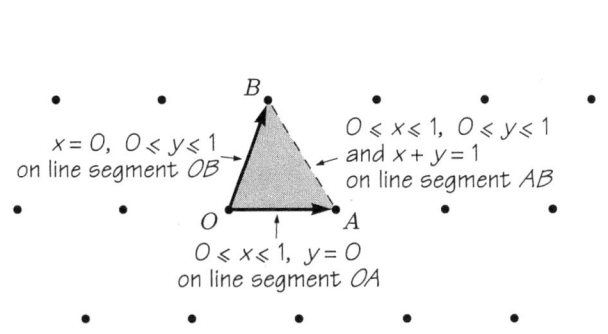

Figure 1.3

Such a point will be in the lattice L if and only if the coefficients x and y are integers. But the only integer solutions to these three inequalities are:

- $x = 0, y = 0$;
- $x = 1, y = 0$;
- $x = 0, y = 1$.

These values correspond to the points O, A and B, respectively.

We have proved that O, A and B are the only lattice points contained in the triangle OAB.

Now let us assume that $\{\mathbf{a}, \mathbf{b}\}$ is a pair of linearly independent vectors in L but that $\{\mathbf{a}, \mathbf{b}\}$ is *not* a basis for L. Then there will be some lattice point P whose vector $\mathbf{p}$ is not an integer linear combination of $\mathbf{a}$ and $\mathbf{b}$ (see Figure 1.4).

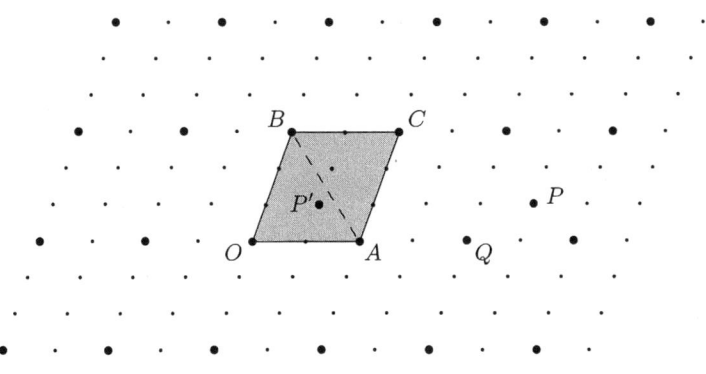

Figure 1.4

We can, however, write $\mathbf{p} = x\mathbf{a} + y\mathbf{b}$, where x and y are some *real* numbers, since $\{\mathbf{a}, \mathbf{b}\}$ is a basis for $\mathbb{R}^2$. Let us put

$$x = n + r, \quad y = m + s,$$

where n and m are integers and where r and s are real numbers, not both zero, satisfying $0 \le r < 1$ and $0 \le s < 1$. The point Q whose vector is $\mathbf{q} = n\mathbf{a} + m\mathbf{b}$ will be a lattice point and, from the conditions on r and s, the point P' whose vector is $\mathbf{p}' = \mathbf{p} - \mathbf{q} = r\mathbf{a} + s\mathbf{b}$ will be a lattice point in the parallelogram $OACB$, where C corresponds to the vector $\mathbf{a} + \mathbf{b}$. If $r + s \le 1$, the point P' must lie in the triangle OAB and will be the point that we are looking for.

If, on the other hand, we have $r + s > 1$ (see Figure 1.5), we can take the lattice point P'' whose vector $\mathbf{p}''$ is given by

$$\mathbf{a} + \mathbf{b} - \mathbf{p}' = (1 - r)\mathbf{a} + (1 - s)\mathbf{b}.$$

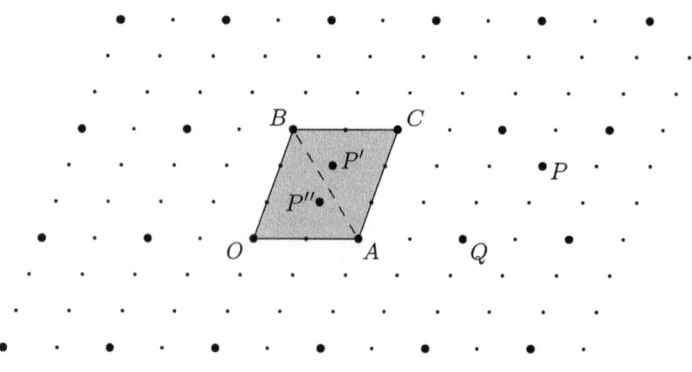

Figure 1.5

Since we have

$$0 < 1 - r, \quad 0 < 1 - s,$$

and

$$(1 - r) + (1 - s) = 2 - (r + s) < 1,$$

it follows that P'' must lie in the triangle OAB. In either case, the points P' or P'' are different from the points O, A and B, since r and s are not both zero and are less than 1.

We have proved that if $\{\mathbf{a}, \mathbf{b}\}$ is not a basis for the lattice L, then the triangle OAB must contain a lattice point different from O, A and B. This means that if the triangle OAB does *not* contain a lattice point other than O, A and B, we can be sure that $\{\mathbf{a}, \mathbf{b}\}$ will be a basis for L. Our proof is complete. ∎

In Exercise 1.2, we saw that a lattice can be expressed as $L(\mathbf{a}, \mathbf{b})$ in a variety of ways using different choices for the basis vectors. Suppose we take another pair of vectors, $\mathbf{a}'$ and $\mathbf{b}'$. It will be useful to know some simple algebraic condition which we could use to test whether or not $\mathbf{a}'$ and $\mathbf{b}'$ will give us the same lattice $L(\mathbf{a}, \mathbf{b})$. We shall investigate this next.

The vectors $\mathbf{a}$ and $\mathbf{b}$ are linearly independent and, if $L(\mathbf{a}', \mathbf{b}')$ is to be a lattice, the vectors $\mathbf{a}'$ and $\mathbf{b}'$ must be linearly independent, too. This implies that we can write

$$\mathbf{a}' = p\mathbf{a} + q\mathbf{b} \quad \text{and} \quad \mathbf{b}' = r\mathbf{a} + s\mathbf{b}$$

and we should be able to solve these equations to obtain $\mathbf{a}$ and $\mathbf{b}$ in terms of $\mathbf{a}'$ and $\mathbf{b}'$. The solutions, as you can check, are

$$\mathbf{a} = (s\mathbf{a}' - q\mathbf{b}')/D \quad \text{and} \quad \mathbf{b} = (-r\mathbf{a}' + p\mathbf{b}')/D$$

where

$$D = ps - qr.$$

It follows that D must be non-zero. But we can say more than this. Since we require $\{\mathbf{a}', \mathbf{b}'\}$ to be a basis for the lattice $L(\mathbf{a}, \mathbf{b})$, both $\mathbf{a}$ and $\mathbf{b}$ must be integer combinations of $\mathbf{a}'$ and $\mathbf{b}'$. This implies that all the coefficients s/D, $-q/D$, $-r/D$ and p/D must be integers. It follows that D must divide the integers p, q, r and s. But then D^2 divides ps and qr, and it follows that D^2 divides $ps - qr = D$. This means that D is either 1 or -1.

With this condition satisfied, all the coefficients above are integers and we can see that not only can a typical element of $L(\mathbf{a}', \mathbf{b}')$ be expressed as a linear combination of $\mathbf{a}$ and $\mathbf{b}$ with integer coefficients, but a typical element of $L(\mathbf{a}, \mathbf{b})$ can be written as a linear combination of $\mathbf{a}'$ and $\mathbf{b}'$, again with integer coefficients. This means that the lattices $L(\mathbf{a}', \mathbf{b}')$ and $L(\mathbf{a}, \mathbf{b})$ are the same.

We have found the condition that we sought. To help you remember it, notice that D is the determinant of the matrix $\begin{bmatrix} p & r \\ q & s \end{bmatrix}$. This is the matrix which converts the coordinates of a vector expressed in terms of $\mathbf{a}'$ and $\mathbf{b}'$ to its coordinates in terms of $\mathbf{a}$ and $\mathbf{b}$. It is known as the **transition matrix** from the basis $\{\mathbf{a}', \mathbf{b}'\}$ to the basis $\{\mathbf{a}, \mathbf{b}\}$. When a vector is written as $x\mathbf{a} + y\mathbf{b}$ and also as $x'\mathbf{a}' + y'\mathbf{b}'$, the coefficients will be related by the equation $\begin{bmatrix} x \\ y \end{bmatrix} = \begin{bmatrix} p & r \\ q & s \end{bmatrix} \begin{bmatrix} x' \\ y' \end{bmatrix}$. For the lattices to be the same, the matrix $\begin{bmatrix} p & r \\ q & s \end{bmatrix}$ must have determinant D equal to 1 or -1. Notice that the inverse of this matrix is the matrix $\begin{bmatrix} s/D & -r/D \\ -q/D & p/D \end{bmatrix}$. This is the transition matrix from the basis $\{\mathbf{a}, \mathbf{b}\}$ to the basis $\{\mathbf{a}', \mathbf{b}'\}$, and its determinant is $1/D$, which must also equal 1 or -1. We have proved the following theorem.

Theorem 1.2 Identity of lattices

Two lattices $L(\mathbf{a}, \mathbf{b})$ and $L(\mathbf{a}', \mathbf{b}')$ are identical if and only if the transition matrix from one basis to the other has integer entries and its determinant is equal to 1 or -1.

Exercise 1.3

The lattice of integer points can be written as $L(\mathbf{a}, \mathbf{b})$ with $\mathbf{a} = (1, 0)$ and $\mathbf{b} = (0, 1)$. Repeat Exercise 1.2 using the condition given in Theorem 1.2. The vectors in (a), (b), (c) and (d) should, in turn, be renamed as $\mathbf{a}'$ and $\mathbf{b}'$ to conform with our discussion above.

Exercise 1.4

Make a sketch of the lattice of integer points. Mark and label as O, A, B, B' and B'' the points represented by the vectors $\mathbf{0} = (0, 0)$, $\mathbf{a} = (1, 0)$, $\mathbf{b} = (0, 1)$, $\mathbf{b}' = (2, 1)$ and $\mathbf{b}'' = (1, 2)$. Also mark and label as C, C' and C'' the points corresponding to $\mathbf{a} + \mathbf{b}$, $\mathbf{a} + \mathbf{b}'$, $\mathbf{a} + \mathbf{b}''$ respectively. Draw the parallelograms $OACB$, $OAC'B'$ and $OAC''B''$. Find the areas of these parallelograms. Which of these parallelograms contain lattice points other than the vertices?

The purpose of Exercise 1.4 was to give you a visual feeling for the concept of a basis for a lattice. The pairs $\{\mathbf{a}, \mathbf{b}\}$ and $\{\mathbf{a}, \mathbf{b}'\}$ are both bases for the lattice L of integer points, but the pair $\{\mathbf{a}, \mathbf{b}''\}$ is not. The areas of the parallelograms $OACB$ and $OAC'B'$ turned out to be the same. This was no coincidence. We shall look into this matter shortly but, before we do, we need to be able to calculate the area of such a parallelogram.

Theorem 1.3

Let P be the area of a parallelogram $OACB$. Let $\mathbf{a} = (a_1, a_2)$ and $\mathbf{b} = (b_1, b_2)$ be the vectors for the points A and B. Then $P^2 = (\mathbf{a} \cdot \mathbf{a})(\mathbf{b} \cdot \mathbf{b}) - (\mathbf{a} \cdot \mathbf{b})^2$ and $P = |a_1 b_2 - a_2 b_1|$.

Proof

Let θ be the angle between $\mathbf{a}$ and $\mathbf{b}$, and let P be the area of $OACB$. Then

$$P = ||\mathbf{a}|| \, ||\mathbf{b}|| \sin \theta. \tag{1.1}$$

Also, by the properties of the dot product,

$$\mathbf{a} \cdot \mathbf{b} = ||\mathbf{a}|| \, ||\mathbf{b}|| \cos \theta. \tag{1.2}$$

See Equation 3.4 in *Unit IB1*.

Squaring and adding Equations 1.1 and 1.2,

$$\begin{aligned} P^2 + (\mathbf{a} \cdot \mathbf{b})^2 &= ||\mathbf{a}||^2 \, ||\mathbf{b}||^2 \sin^2 \theta + ||\mathbf{a}||^2 \, ||\mathbf{b}||^2 \cos^2 \theta \\ &= ||\mathbf{a}||^2 \, ||\mathbf{b}||^2 \quad (\text{since } \sin^2 \theta + \cos^2 \theta = 1) \\ &= (\mathbf{a} \cdot \mathbf{a})(\mathbf{b} \cdot \mathbf{b}). \end{aligned}$$

Thus,

$$P^2 = (\mathbf{a} \cdot \mathbf{a})(\mathbf{b} \cdot \mathbf{b}) - (\mathbf{a} \cdot \mathbf{b})^2,$$

as required.

Working this out in detail,

$$P^2 = (a_1^2 + a_2^2)(b_1^2 + b_2^2) - (a_1 b_1 + a_2 b_2)^2,$$

which, as you may check, simplifies to

$$P^2 = (a_1 b_2 - a_2 b_1)^2,$$

and, on taking square roots:

$P = |a_1 b_2 - a_2 b_1|$, as required. ∎

Exercise 1.5

Let O, A, B, and C be the points in the plane which are represented by vectors $\mathbf{0}, \mathbf{a}, \mathbf{b}$, and $\mathbf{a} + \mathbf{b}$. Find the area of the parallelogram $OACB$ in the following cases.

(a) $\mathbf{a} = (1, 1)$, $\mathbf{b} = (-1, 5)$
(b) $\mathbf{a} = (1, \sqrt{3})$, $\mathbf{b} = (\sqrt{3}, -1)$
(c) $\mathbf{a} = (-\sqrt{3}, 1)$, $\mathbf{b} = (\sqrt{3}, 1)$

We shall find it useful to have available the term *basic parallelogram*, so we define it now.

Definition 1.3 Basic parallelogram

If $\{\mathbf{a}, \mathbf{b}\}$ is a basis for a lattice L and if O, A, B and C are the points of L with vectors $\mathbf{0}, \mathbf{a}, \mathbf{b}$ and $\mathbf{a} + \mathbf{b}$, then the parallelogram $OACB$ is called a **basic parallelogram** for L.

It is sometimes useful to relate the concept of a lattice to that of a tiling. Suppose we take some basic parallelogram for a lattice L. We can obtain a tiling of the plane by applying to this parallelogram all the translations of the lattice. The lattice points L are the set of vertices of the tiling. There will be any number of such tilings of the plane associated with a given lattice, since we can make any number of choices for the basic parallelogram.

You saw two examples of basic parallelograms in Exercise 1.4 and we remarked that they had the same area. This is true in general, as we shall see from the following theorem.

> **Theorem 1.4**
>
> Let $OACB$ be a basic parallelogram for a lattice L. Then any parallelogram in L with one vertex at O will be basic if and only if its area is the same as that of $OACB$.

When we speak of a 'parallelogram in a lattice', we are referring to a parallelogram whose vertices are points of the lattice.

Proof

Let $OA'C'B'$ be a parallelogram in the lattice L, and let $\mathbf{a}' = (a_1', a_2')$ and $\mathbf{b}' = (b_1', b_2')$ be the vectors corresponding to A' and B'. We can then write $\mathbf{a}' = p\mathbf{a} + q\mathbf{b}$ and $\mathbf{b}' = r\mathbf{a} + s\mathbf{b}$, where p, q, r and s are integers. That is to say,

$$a_1' = pa_1 + qb_1, \quad a_2' = pa_2 + qb_2,$$
$$b_1' = ra_1 + sb_1, \quad b_2' = ra_2 + sb_2.$$

In matrix form, these equations are

$$\begin{bmatrix} a_1' & a_2' \\ b_1' & b_2' \end{bmatrix} = \begin{bmatrix} p & q \\ r & s \end{bmatrix} \begin{bmatrix} a_1 & a_2 \\ b_1 & b_2 \end{bmatrix}.$$

Forming the determinant of both sides, we get the equation

$$(a_1' b_2' - a_2' b_1') = (ps - qr)(a_1 b_2 - a_2 b_1).$$

$\det(AB) = \det A \det B$

Taking absolute values, we have

$$|a_1' b_2' - a_2' b_1'| = |ps - qr| \, |a_1 b_2 - a_2 b_1|.$$

If $OA'C'B'$ is a basic parallelogram then, from Theorem 1.2, we know that $ps - qr$ will equal 1 or -1, and so $|a_1' b_2' - a_2' b_1'| = |a_1 b_2 - a_2 b_1|$. Hence, by Theorem 1.3, the areas of $OA'C'B'$ and $OACB$ will be the same. Conversely, if the areas of $OA'C'B'$ and $OACB$ are the same, then $|ps - qr|$ will equal 1 and so $ps - qr$ will be either 1 or -1. It follows, again from Theorem 1.2, that $OA'C'B'$ is a basic parallelogram. ∎

Exercise 1.6

Show that the area of any parallelogram whose vertices are lattice points will be an integer multiple of the area of a basic parallelogram.

1.2 The minimality conditions in the plane

Given a lattice L, there are, as you have seen, many ways of choosing a basis for L. You may ask whether any one basis is better than another. In some situations it doesn't matter which basis you take, but there are occasions where we find it useful to have one which satisfies what we call the *minimality conditions*. Before we discuss these conditions, let us show that any disc will contain only a finite number of lattice points. This may seem intuitively obvious, but let us see how we can formally establish this fact.

> **Theorem 1.5**
>
> For any lattice L and any disc D in the plane, there are only finitely many lattice points in the disc.

Proof

Let $\{\mathbf{a}, \mathbf{b}\}$ be a basis for L. Then every lattice point is represented by a vector of the form $n\mathbf{a} + m\mathbf{b}$, where n and m are integers.

Consider the line of points obtained by letting n vary while keeping m fixed. That is, for each $m \in \mathbb{Z}$, let

$$\mathcal{L}_m = \{n\mathbf{a} + m\mathbf{b} : n \in \mathbb{Z}\}.$$

Successive points of $\mathcal{L}_m$ are separated by a distance $\|\mathbf{a}\|$, and so, for any m, there is only a finite number of points of $\mathcal{L}_m$ in any disc D. (If D has diameter d, then the number of such points cannot exceed $\dfrac{d}{\|\mathbf{a}\|} + 1$.)

Now two successive lines $\mathcal{L}_m$ and $\mathcal{L}_{m+1}$ are separated by a perpendicular distance $\dfrac{|a_1 b_2 - a_2 b_1|}{\|\mathbf{a}\|}$, since $|a_1 b_2 - a_2 b_1|$ is the area of a basic parallelogram (see Figure 1.6).

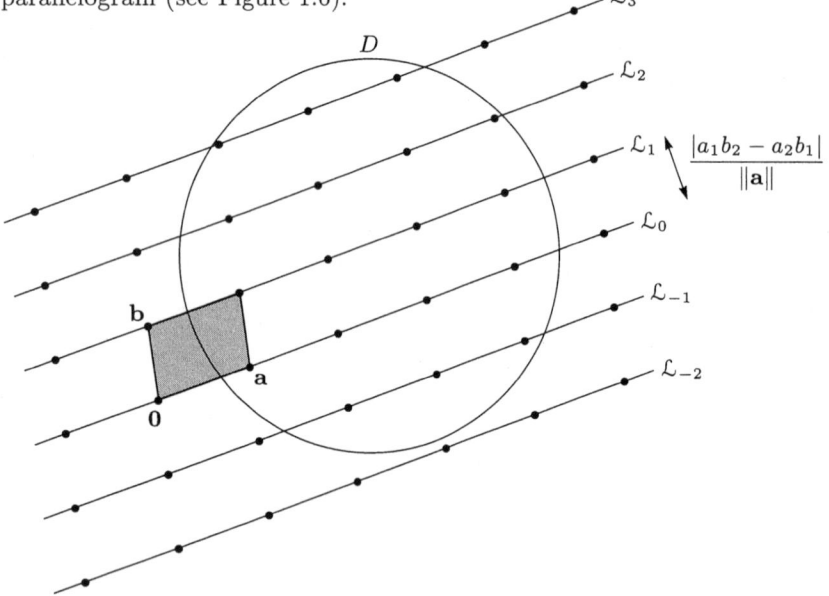

Figure 1.6

Therefore, only a finite number of the lines $\mathcal{L}_m$ will intersect any disc D. (If D has diameter d, then the number of such lines cannot exceed $\dfrac{d\|\mathbf{a}\|}{|a_1 b_2 - a_2 b_1|} + 1$.)

But we have seen that any $\mathcal{L}_m$ which does intersect D has only a finite number of points in D. Therefore, the total number of lattice points in D must be finite. ∎

Exercise 1.7

Let $L = L(\mathbf{a}, \mathbf{b})$ where $\mathbf{a} = (3, 4), \mathbf{b} = (4, 3)$. Find a number N such that there cannot be more than N lattice points in any disc D of diameter 10.

The number found in this way is usually a considerable overestimate. As a matter of interest, in the case of the lattice and disc of Exercise 1.7, the seven lines $\mathcal{L}_{-3}, \mathcal{L}_{-2}, \ldots, \mathcal{L}_3$ intersect D; $\mathcal{L}_0$ intersects D in the three points $(-3, -4), (0, 0), (3, 4)$; $\mathcal{L}_1$ and $\mathcal{L}_{-1}$ intersect D in two points each; and the other lines intersect D in just one point each. Thus there are eleven points of L in the disc altogether (see Figure 1.7).

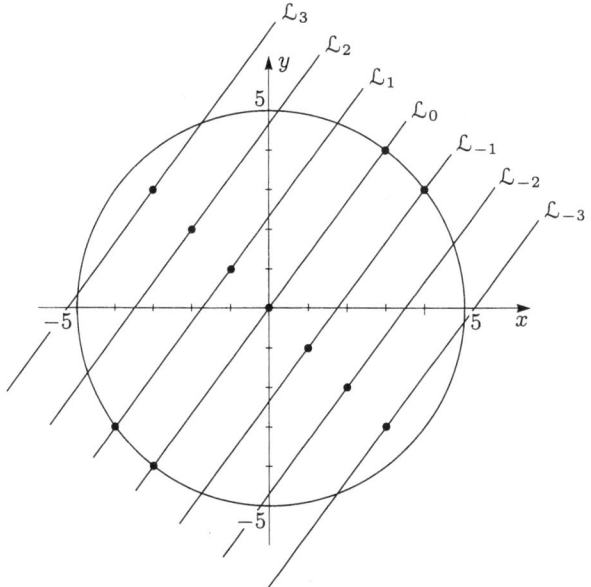

Figure 1.7

We now discuss the *minimality conditions*.

Definition 1.4 Minimality conditions

An ordered pair of vectors $\{\mathbf{a}, \mathbf{b}\}$ in a lattice L is said to satisfy the **minimality conditions** if:

(a) of the set of all non-zero vectors in L, the vector $\mathbf{a}$ has least magnitude;

(b) of the set of all vectors in L which are not multiples of $\mathbf{a}$, the vector $\mathbf{b}$ has least magnitude.

It is not clear from the definition above that a pair of vectors $\{\mathbf{a}, \mathbf{b}\}$ satisfying these conditions will always exist. We need to see that, of every set of vectors representing points of a lattice, there must be at least one which has least magnitude. In the case of a finite set of vectors, there is no difficulty, but the sets that we are interested in are infinite. The way round the problem is to look only at the lattice points inside some disc with centre O. The corresponding vectors clearly have smaller magnitude than those outside the disc, and we have shown in Theorem 1.5 that there can be only a finite number of lattice points in a disc. This assures us that a suitable pair $\{\mathbf{a}, \mathbf{b}\}$ exists, and it gives us the following strategy for finding $\mathbf{a}$ and $\mathbf{b}$.

(a) Choose a disc about O which contains a basis for the lattice.

(b) From the finite set of lattice points in the disc, choose $\mathbf{a}$ to be a vector which has least magnitude.

(c) From the remaining lattice points in the disc which are not multiples of $\mathbf{a}$, choose $\mathbf{b}$ to be a vector with least magnitude.

If the pair of vectors $\{\mathbf{a}, \mathbf{b}\}$ satisfies the minimality conditions, then it is easy to see that the pairs $\{-\mathbf{a}, \mathbf{b}\}$, $\{\mathbf{a}, -\mathbf{b}\}$ and $\{-\mathbf{a}, -\mathbf{b}\}$ must also satisfy the conditions. There may also be other pairs, as you will see in later exercises.

An ordered pair of vectors $\{\mathbf{a}, \mathbf{b}\}$ which satisfies these two conditions will be linearly independent, since neither $\mathbf{a}$ nor $\mathbf{b}$ is the zero vector and $\mathbf{b}$ is not a multiple of $\mathbf{a}$. We shall prove very shortly that such a pair of vectors $\{\mathbf{a}, \mathbf{b}\}$ must be a basis for L. It is called a *reduced basis* for L.

Before we consider this, let us remind you of an important property concerning the magnitude of vectors.

Suppose that **u** and **v** are any two vectors in $\mathbb{R}^2$. Let us consider the magnitude of the vector $\|\mathbf{u}+\mathbf{v}\|$. We may write

$$\|\mathbf{u}+\mathbf{v}\|^2 = (\mathbf{u}+\mathbf{v})\cdot(\mathbf{u}+\mathbf{v})$$
$$= \mathbf{u}\cdot\mathbf{u} + 2\mathbf{u}\cdot\mathbf{v} + \mathbf{v}\cdot\mathbf{v}$$
$$= \|\mathbf{u}\|^2 + 2\mathbf{u}\cdot\mathbf{v} + \|\mathbf{v}\|^2.$$

But $\mathbf{u}\cdot\mathbf{v} = \|\mathbf{u}\|\,\|\mathbf{v}\|\cos\theta$, where θ is the angle between the two vectors and, since $\cos\theta \leq 1$, it follows that $\mathbf{u}\cdot\mathbf{v} \leq \|\mathbf{u}\|\,\|\mathbf{v}\|$.

From this inequality, we obtain

$$\|\mathbf{u}+\mathbf{v}\|^2 \leq \|\mathbf{u}\|^2 + 2\|\mathbf{u}\|\,\|\mathbf{v}\| + \|\mathbf{v}\|^2 = (\|\mathbf{u}\| + \|\mathbf{v}\|)^2.$$

Then, taking the square root of each side, we find that

$$\|\mathbf{u}+\mathbf{v}\| \leq \|\mathbf{u}\| + \|\mathbf{v}\|.$$

This inequality is called the **triangle inequality**, and it holds for all vectors in $\mathbb{R}^2$.

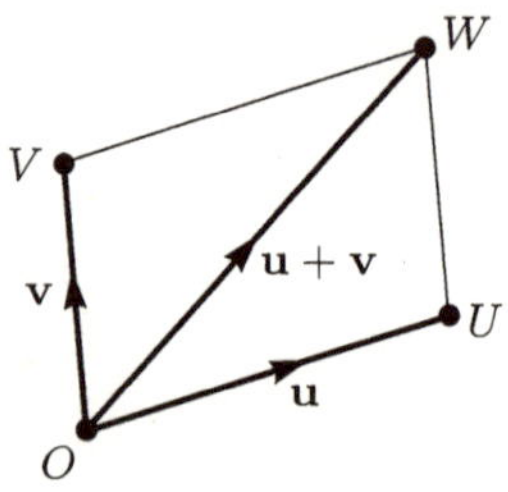

When the vectors **u** and **v** are linearly independent, then $\cos\theta < 1$, and so we get the **strict triangle inequality**

$$\|\mathbf{u}+\mathbf{v}\| < \|\mathbf{u}\| + \|\mathbf{v}\|.$$

To understand this geometrically, take points O, U, V and W whose vectors are $\mathbf{0}, \mathbf{u}, \mathbf{v}$ and $\mathbf{u}+\mathbf{v}$. They will form a parallelogram $OUWV$ (see Figure 1.8). In the triangle OUW, the lengths of the line segments OU and OW are $\|\mathbf{u}\|$ and $\|\mathbf{u}+\mathbf{v}\|$, and the length of the line segment UW equals that of OV, which is $\|\mathbf{v}\|$. The inequality states

length of OW < length of OU + length of UW.

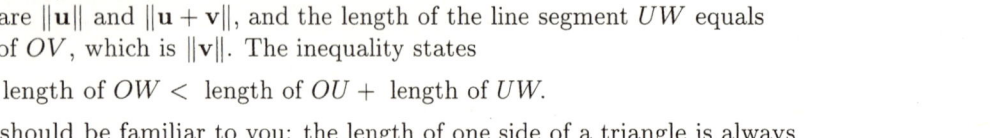

Figure 1.8

This should be familiar to you: the length of one side of a triangle is always less than the sum of the lengths of the other two sides.

We now present a theorem which will be very useful in our later investigations, both in this unit and in *Unit GE4*.

Theorem 1.6

Let L be a plane lattice, and let $\{\mathbf{a}, \mathbf{b}\}$ be a pair of vectors in L which satisfies the minimality conditions. Let θ be the angle between **a** and **b**. Then:

(a) $\|\mathbf{a}\| \leq \|\mathbf{b}\|$;
(b) $-\|\mathbf{a}\|/2 \leq \|\mathbf{b}\|\cos\theta \leq \|\mathbf{a}\|/2$;
(c) $\pi/3 \leq \theta \leq 2\pi/3$;
(d) $\{\mathbf{a}, \mathbf{b}\}$ is a basis for L.

When we speak of the angle between two vectors, we mean an angle which lies between 0 and π.

Proof

(a) The vector **a** was chosen to have least magnitude $\|\mathbf{a}\|$ from the set of all lattice vectors other than **0**, whereas **b** was chosen to have least magnitude $\|\mathbf{b}\|$ from a subset of these. It follows that (a) holds.

(b) To show that (b) holds, we consider first the case where $\cos\theta \geq 0$. In Figure 1.9, A and B are the lattice points whose position vectors are **a** and **b**, respectively. In the triangle OAB, the distance from B to A is equal to $\|\mathbf{a}-\mathbf{b}\|$ and the distance from B to O is $\|\mathbf{b}\|$. Since the vector $\mathbf{a}-\mathbf{b}$ is not a multiple of **a**, the second of the minimality conditions gives us $\|\mathbf{b}\| \leq \|\mathbf{a}-\mathbf{b}\|$. This means that the point B must lie either on or to the left of the perpendicular bisector of OA. It follows that

$$\|\mathbf{b}\|\cos\theta \leq \|\mathbf{a}\|/2.$$

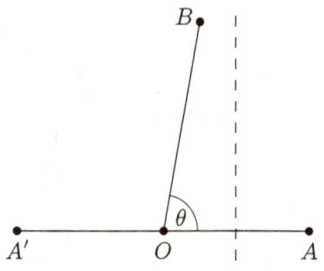

Figure 1.9

Now for the case where $\cos\theta < 0$. In Figure 1.10, A' and B are lattice points whose position vectors are $-\mathbf{a}$ and **b**. The distance from B to A'

is equal to $\|\mathbf{a}+\mathbf{b}\|$ and the distance from B to O is $\|\mathbf{b}\|$. Using the fact that $\mathbf{a}+\mathbf{b}$ is not a multiple of $\mathbf{a}$, it follows that $\|\mathbf{b}\| \leq \|\mathbf{a}+\mathbf{b}\|$, and this means that B must lie on or to the right of the perpendicular bisector of OA'. Hence

$$-\|\mathbf{a}\|/2 \leq \|\mathbf{b}\| \cos\theta.$$

In both cases we get

$$-\|\mathbf{a}\|/2 \leq \|\mathbf{b}\| \cos\theta \leq \|\mathbf{a}\|/2,$$

so (b) holds.

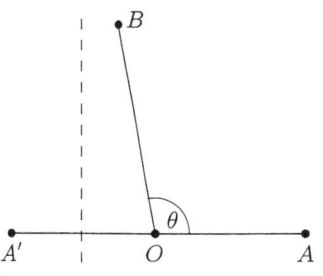

Figure 1.10

(c) Using the inequalities in (a) and (b), we obtain the inequalities

$$-\|\mathbf{b}\|/2 \leq \|\mathbf{b}\| \cos\theta \leq \|\mathbf{b}\|/2,$$

from which we get

$$-1/2 \leq \cos\theta \leq 1/2.$$

It follows that

$$\pi/3 \leq \theta \leq 2\pi/3,$$

so (c) holds.

(d) Finally we prove (d). Suppose that $\{\mathbf{a}, \mathbf{b}\}$ is *not* a basis for L. By Theorem 1.1, we know that the triangle OAB must contain a lattice point P different from O, A and B. We can write the corresponding vector $\mathbf{p}$ as $x\mathbf{a} + y\mathbf{b}$, where we have the inequalities $0 \leq x$, $0 \leq y$ and $x + y \leq 1$. Since $\mathbf{p}$ is not one of the vectors $\mathbf{a}$, $\mathbf{b}$ or $\mathbf{0}$, it follows that $x < 1$, $y < 1$, and that x and y cannot both be zero.

If $y = 0$, then $\mathbf{p} = x\mathbf{a}$ is a non-zero vector such that $\|\mathbf{p}\| < \|\mathbf{a}\|$, and this contradicts the first minimality condition. If $x = 0$, then $\mathbf{p} = y\mathbf{b}$ and we have $\|\mathbf{p}\| < \|\mathbf{b}\|$, which contradicts the second minimality condition since $\mathbf{p}$ is not a multiple of $\mathbf{a}$.

The remaining case is where x and y are both non-zero. Here the vectors $x\mathbf{a}$ and $y\mathbf{b}$ are linearly independent, and the strict triangle inequality gives us

$$\|\mathbf{p}\| = \|x\mathbf{a} + y\mathbf{b}\| < \|x\mathbf{a}\| + \|y\mathbf{b}\| = x\|\mathbf{a}\| + y\|\mathbf{b}\|.$$

But $\|\mathbf{a}\| \leq \|\mathbf{b}\|$, so we see that

$$x\|\mathbf{a}\| + y\|\mathbf{b}\| \leq (x+y)\|\mathbf{b}\| \leq \|\mathbf{b}\|.$$

Hence $\|\mathbf{p}\| < \|\mathbf{b}\|$, and this also contradicts the second minimality condition.

In all cases, we obtain a contradiction, so $\{\mathbf{a}, \mathbf{b}\}$ must be a basis for the lattice L. The proof of Theorem 1.6 is complete. ∎

Definition 1.5 Reduced basis

A pair of vectors $\{\mathbf{a}, \mathbf{b}\}$ in some lattice L is called a **reduced basis** for L if the pair $\{\mathbf{a}, \mathbf{b}\}$ satisfies the minimality conditions.

You should note that, when we say that $\{\mathbf{a}, \mathbf{b}\}$ is a reduced basis for a lattice L, we are speaking about an ordered pair of vectors. The first vector always has the least magnitude.

Exercise 1.8

Find a reduced basis for the lattice $L((3,2), (3,4))$.

Exercise 1.9

Find a reduced basis for the lattice $L((2,1), (2,-1))$.

This completes the basic preparation for our investigation of lattices.

2 SYMMETRIES OF A PLANE LATTICE

In this section, we examine the various types of symmetry that a lattice may possess. For a given lattice, we show that only certain translational symmetries will be present and that the possible angles for a rotational symmetry of a lattice are very restricted.

The material covered in this section applies to any lattice. When we come to study particular lattices in later sections of this unit, we shall find that not all the symmetries described here occur in all lattices. Indeed, rotational symmetries of orders 3 and 4 (for example) *cannot* occur in the same plane lattice.

2.1 Types of symmetry

The symmetries of a lattice consist of those isometries of the plane which map the lattice to itself. From Theorem 5.1 of *Unit IB1*, you will know that there are six types of plane isometry, namely the identity, non-zero translations, rotations through π, rotations through non-multiples of π, reflections and glide reflections.

Every lattice will possess, among its symmetries, the identity e which maps every lattice point to itself. Because of its unique properties, we classify it on its own. We do, however, often wish to include it among the translations and write it as $t[\mathbf{0}]$. It can also be regarded as a rotation about any point with angle 0 or some other multiple of 2π. There is little more to be said about it, so we now turn to the translations.

A translation of $\mathbb{R}^2$, you will recall, is an isometry $t[\mathbf{p}]$ which maps a vector $\mathbf{x}$ in $\mathbb{R}^2$ to a vector $\mathbf{x} + \mathbf{p}$ in $\mathbb{R}^2$. There is an obvious one–one correspondence between the translations $t[\mathbf{p}]$ and the vectors $\mathbf{p}$ in $\mathbb{R}^2$, which is given by $t[\mathbf{p}] \mapsto t[\mathbf{p}](\mathbf{0}) = \mathbf{p}$. By the **magnitude** of a translation $t[\mathbf{p}]$ we mean the magnitude $\|\mathbf{p}\|$ of the corresponding vector $\mathbf{p}$. We shall say that a translation is **proper** if its magnitude is non-zero. Two proper translations $t[\mathbf{p}]$ and $t[\mathbf{q}]$ are said to be **parallel** when the corresponding vectors $\mathbf{p}$ and $\mathbf{q}$ have the same direction.

In the context of lattices, we shall be concerned mainly with those translations of $\mathbb{R}^2$ which correspond to vectors representing points of the lattice. For a translation $t[\mathbf{p}]$ to be a symmetry of a lattice L, it is both necessary and sufficient that the point $\mathbf{p}$ be one of the lattice points. Such a translation will be called a **translation of the lattice**. If $\{\mathbf{a}, \mathbf{b}\}$ is a basis for a lattice L, then the translations of the lattice L are all of the form $(t[\mathbf{a}])^n (t[\mathbf{b}])^m$, where n and m are integers. Clearly, a translation written in this form maps the origin O to the point in the plane whose vector is $n\mathbf{a} + m\mathbf{b}$. The set of all translations of the lattice L forms the group

$$T = \{(t[\mathbf{a}])^n (t[\mathbf{b}])^m : n, m \in \mathbb{Z},\ t[\mathbf{b}]\,t[\mathbf{a}] = t[\mathbf{a}]\,t[\mathbf{b}]\}.$$

In Section 4 we introduce the subscript 2, and call this group T_2 (see page 37).

Thus, $t[\mathbf{a}]$ and $t[\mathbf{b}]$ generate T, in the sense that was introduced in Section 3 of *Unit IB2*.

The translations $t[\mathbf{a}]$ and $t[\mathbf{b}]$ are not the only possible generators of the group T. If $\{\mathbf{a}', \mathbf{b}'\}$ is another basis for the lattice L, then the corresponding translations $t[\mathbf{a}']$ and $t[\mathbf{b}']$ would serve equally well as a pair of generators for T.

A rotation of $\mathbb{R}^2$ which is a symmetry of a lattice L will be called a **rotation of the lattice**. There is certainly one rotation of $\mathbb{R}^2$ other than e which is a symmetry of every lattice. Let us take the mapping which sends a point $\mathbf{x}$ to the point $-\mathbf{x}$. This is the rotation $r[\pi]$ with centre at the origin and angle π. If we take any point $n\mathbf{a} + m\mathbf{b}$ in a lattice $L = L(\mathbf{a}, \mathbf{b})$, this

In three dimensions, the mapping $\mathbf{x} \mapsto -\mathbf{x}$ is not a rotation. It is called *central inversion*.

mapping will send it to the point $-n\mathbf{a} - m\mathbf{b}$, which is also a point of the lattice L. The mapping is therefore a symmetry of L.

Suppose that r is some rotation of a lattice L and that its angle is θ. If some multiple of θ is a multiple of 2π, there will be a least positive integer n for which $n\theta$ is a multiple of 2π. This number is the least positive integer for which $r^n = e$. In other words, it is the order of the rotation r. You can see that the rotation $r[\pi]$ which we have just been discussing is a rotation of order 2. Later in this section, we shall prove that the order of a rotation of a lattice can take only certain values.

There are two further types of isometries which may occur as symmetries of a lattice. These are reflections and glide reflections. When a reflection of $\mathbb{R}^2$ is a symmetry of a lattice L, we say that it is a **reflection of the lattice**. Similarly, we say that a glide reflection of $\mathbb{R}^2$ is a **glide reflection of the lattice** if it is a symmetry of L. We shall see that not every lattice has reflections or glide reflections among its symmetries.

Every reflection of $\mathbb{R}^2$ is an element of order 2, since performing it twice restores each point to its original position. In algebraic terms this means that $q^2 = e$ whenever q is a reflection. Suppose we now take some glide reflection g of a lattice L. We know that we can write it in the notation of Section 5 of *Unit IB1*, as

$$g = q[\mathbf{g}, \mathbf{c}, \theta] = t[\mathbf{g}] \, q[\mathbf{c}, \theta]$$

where $t[\mathbf{g}]$ is a translation parallel to the axis of the reflection $q[\mathbf{c}, \theta]$.

Definition 2.1

When a glide reflection is written in the form

$$g = q[\mathbf{g}, \mathbf{c}, \theta] = t[\mathbf{g}] \, q[\mathbf{c}, \theta],$$

then $t[\mathbf{g}]$ and $q[\mathbf{c}, \theta]$ are said to be the **translation component** and the **reflection component** of g.

Note carefully that the translation and reflection *components* of a glide reflection g are **not** in general the same as the translation *part* and the linear *part* obtained when you express g in its standard form! You can see this if you consult Equation 14 of the Isometry Toolkit, where $q[\mathbf{g}, \mathbf{c}, \theta]$ is expressed in standard form as

$$q[\mathbf{g}, \mathbf{c}, \theta] = t[\mathbf{d}] \, q[\theta], \quad \text{where } \mathbf{d} = \mathbf{g} + \mathbf{c} - q[\theta](\mathbf{c}).$$

Thus, the translation *component* of $q[\mathbf{g}, \mathbf{c}, \theta]$ is $t[\mathbf{g}]$ while the translation *part* is $t[\mathbf{d}]$. Also, the reflection *component* is $q[\mathbf{c}, \theta]$ while the linear *part* is $q[\theta]$.

Exercise 2.1

Find the translation component, the translation part, the reflection component and the linear part of each of the following glide reflections.

(a) $g = q[(2,0), (0,1), 0]$

(b) $g = q[(0,2), (0,1), \pi/2]$

(c) $g = q[(3,3), (2,0), \pi/4]$

This distinction between components and parts is a very important one: if g is a glide reflection of a lattice L, then its translation and reflection *components* are *not* necessarily symmetries of L, but its translation and linear *parts* are *always* symmetries of L.

Example 2.1

Figure 2.1 shows the lattice $L = L(\mathbf{a}, \mathbf{b})$ where $\mathbf{a} = (2, -1)$ and $\mathbf{b} = (2, 1)$. Let $g = q[(2,0), (0, \frac{1}{2}), 0], h = q[(4,0), (0,1), 0]$.

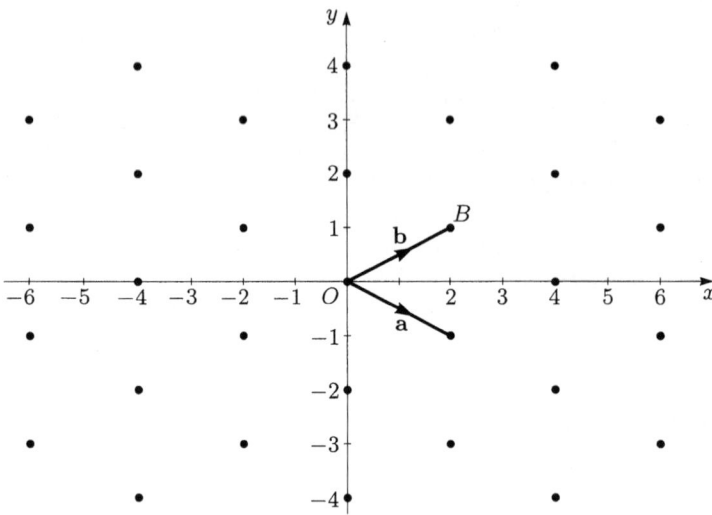

Figure 2.1

Since g maps horizontal lines to horizontal lines, and the point O to the point B, it follows that g maps the lattice points on the x-axis to those on the line $y = 1$. As one horizontal line of points is mapped to another, it is geometrically clear that $g(L) = L$, so that g is indeed a symmetry of L. However, the translation component of g is $t[(2,0)]$, which maps O to $(2,0)$ — not a point in L. Similarly, the reflection component maps O to $(0,1)$, which is again not a point in L. Thus the translation and reflection components of g are not symmetries of L.

When we look at h, however, the story is different. It is easy to see geometrically that $h(L) = L$. Also, the translation component is $t[(4,0)]$, and since this translates O to a point of L, it translates every point of L to another point of L, and is a translation of L. The reflection component consists of reflection in a horizontal line of lattice points, and it is easy to see geometrically that it is a reflection of L.   ♦

Thus, a lattice may have glide reflections of two distinct types. We call these *essential* and *inessential*.

Definition 2.2 Essential and inessential glide reflections

Let $g = q[\mathbf{g}, \mathbf{c}, \theta]$ be a glide reflection of a lattice L, so that $g = t[\mathbf{g}]\, q[\mathbf{c}, \theta]$ where $\mathbf{g}$ is in the direction of the axis of $q[\mathbf{c}, \theta]$. If neither $t[\mathbf{g}]$ nor $q[\mathbf{c}, \theta]$ are symmetries of L, then g is an **essential glide reflection** of L. If $t[\mathbf{g}]$ and $q[\mathbf{c}, \theta]$ *are* symmetries of L, then g is an **inessential glide reflection** of L.

Exercise 2.2

Let L be the lattice $L(\mathbf{a}, \mathbf{b})$ of Example 2.1. One of the following glide reflections of L is essential and one is inessential. Which is which?
(a) $g = q[(0,2), (0,0), \pi/2]$
(b) $g = q[(0,1), (1,0), \pi/2]$

We have not yet proved that the translational and linear *parts* of a symmetry of a lattice L are themselves symmetries of L. This is one of the results of the next subsection.

2.2 Symmetries which fix O

As we know from *Unit IB1*, the isometries that fix the origin O are the rotations with centre O and reflections in lines passing through O. The rules for composing such isometries were established in Section 5 of *Unit IB1* and are to be found in the Isometry Toolkit.

Let L be a plane lattice. We shall now consider the subgroup of $\Gamma(L)$ consisting of those symmetries of L that fix O. We shall denote this subgroup by $\Gamma_O(L)$.

By Theorem 1.5, there are only finitely many lattice points in any disc. Let us choose such a disc, with centre O and containing some lattice points other than O, and let X be the set of lattice points in the disc. The group $\Gamma_O(L)$ must clearly act on the set X. Let $\mathbf{x}$ be a point in X other than O and consider the stabilizer $\text{Stab}(\mathbf{x})$. It contains no non-zero rotations and at most one reflection, i.e. the one in the line through O and $\mathbf{x}$. Hence $|\text{Stab}(\mathbf{x})| \leq 2$. Now $\text{Orb}(\mathbf{x})$ is finite and $|\Gamma_O(L)| = |\text{Stab}(\mathbf{x})|\,|\text{Orb}(\mathbf{x})|$, so $\Gamma_O(L)$ must be a finite group.

The stabilizer $\text{Stab}(\mathbf{0})$ *is, of course, the whole of* $\Gamma_O(L)$.

Remember the Orbit–stabilizer Theorem in Unit GE1!

As a consequence, there is only a finite number of rotations in $\Gamma_O(L)$, and each has finite order. Hence there is some maximum order n that such a rotation can have. Let $r[\theta]$ have this maximum order n. The powers of $r[\theta]$ form a cyclic group of order n. Now this group is also generated by the rotation $r[2\pi/n]$. This implies that $r[2\pi/n]$ must be a rotation of the lattice, and that every rotation of the lattice which fixes O must be a power of $r[2\pi/n]$.

It will be useful to have a way of telling whether or not some isometry which fixes O is a symmetry of a lattice $L(\mathbf{a}, \mathbf{b})$. The next theorem gives this condition. The proof makes use of the fact that isometries that fix the origin preserve dot products. This was proved in Section 3 of *Unit IB1*.

Theorem 2.1

Let L be a lattice with basis $\{\mathbf{a}, \mathbf{b}\}$ and let f be a plane isometry such that $f(\mathbf{0}) = \mathbf{0}$. Then f is a symmetry of L if and only if $\{f(\mathbf{a}), f(\mathbf{b})\}$ is also a basis for L.

Proof

The parallelogram $OACB$, where the vertices A and B have vectors $\mathbf{a}$ and $\mathbf{b}$, is a basic parallelogram. Since f is linear, the image of $OACB$ under f is the parallelogram whose vertices are O, A', C' and B', where A' and B' have vectors $f(\mathbf{a})$ and $f(\mathbf{b})$. If P is the area of $OACB$ then, by Theorem 1.3, we have

$$P^2 = (\mathbf{a} \cdot \mathbf{a})(\mathbf{b} \cdot \mathbf{b}) - (\mathbf{a} \cdot \mathbf{b})^2.$$

Since f is an isometry which fixes $\mathbf{0}$, it preserves dot products, so the areas of $OA'C'B'$ and $OACB$ must be the same.

Suppose that f is a symmetry of L. Then $f(\mathbf{a})$ and $f(\mathbf{b})$ are points of L, and so $OA'C'B'$ is a parallelogram of the lattice L. By Theorem 1.4, the pair $\{f(\mathbf{a}), f(\mathbf{b})\}$ is a basis for L.

Conversely, suppose that $\{f(\mathbf{a}), f(\mathbf{b})\}$ is a basis for L. Then certainly, both $f(\mathbf{a})$ and $f(\mathbf{b})$ are points of the lattice L. For any point $n\mathbf{a} + m\mathbf{b}$ in L, its image under f is $f(n\mathbf{a} + m\mathbf{b}) = nf(\mathbf{a}) + mf(\mathbf{b})$, since f is linear, and this point is also in L. It follows that f is a symmetry of L. ∎

2.3 Composites of symmetries

In this subsection, we shall examine composites of the symmetries of a lattice. As every symmetry of a lattice is a plane isometry, you will find the Isometry Toolkit useful in performing the relevant calculations.

Recall that, when a plane isometry f is written in standard form, as

$$f = t[\mathbf{p}]\, r[\theta] \quad \text{or} \quad f = t[\mathbf{p}]\, q[\theta],$$

the isometries $t[\mathbf{p}]$ and $r[\theta]$ or $q[\theta]$ are called the *translation part* and the *linear part* respectively. We shall now prove the result which we mentioned in the previous subsection.

> **Theorem 2.2**
>
> Let f be a symmetry of a plane lattice L. Then the translation and linear parts of f are themselves symmetries of L.

Proof

Let $f = t[\mathbf{p}]\, r[\theta]$ or $t[\mathbf{p}]\, q[\theta]$. The point $\mathbf{0}$ belongs to L and, since f is a symmetry of L, it follows that $f(\mathbf{0}) = \mathbf{p}$ also belongs to L. Thus, $t[\mathbf{p}]$ is a symmetry of L.

Since the symmetries of L form a group, $t[-\mathbf{p}]$ is also a symmetry, and so, therefore, is $t[-\mathbf{p}]\, f$. But this is just the linear part of f (i.e. $r[\theta]$ or $q[\theta]$), which is therefore also a symmetry of L. ∎

In *Unit IB1*, you saw how every composite of isometries may be expressed in standard form. We can do precisely the same with the symmetries of a lattice. The same formulae apply, and these are given in the Isometry Toolkit. To give you some practice with these formulae, here are some exercises.

Exercise 2.3

Let L be the lattice $L(\mathbf{a}, \mathbf{b})$, where $\mathbf{a} = (1,0)$ and $\mathbf{b} = (0,1)$. Find the vectors $r[\pi/2](\mathbf{a})$, $r[\pi/2](\mathbf{b})$, $q[\pi/4](\mathbf{a})$ and $q[\pi/4](\mathbf{b})$, expressing your answers as linear combinations of $\mathbf{a}$ and $\mathbf{b}$. Then calculate the composite $f_1 f_2$ in the following cases, expressing your result in standard form.

(a) $f_1 = r[\pi/2]$, $f_2 = t[\mathbf{a} - \mathbf{b}]\, r[\pi/2]$
(b) $f_1 = q[\pi/4]$, $f_2 = t[\mathbf{a} + \mathbf{b}]\, q[3\pi/4]$
(c) $f_1 = t[\mathbf{a}]\, r[\pi/2]$, $f_2 = t[\mathbf{b}]\, q[0]$
(d) $f_1 = t[\mathbf{a}]\, q[\pi/4]$, $f_2 = t[\mathbf{a} - \mathbf{b}]\, r[\pi/2]$

Exercise 2.4

Let L be the lattice $L(\mathbf{a}, \mathbf{b})$, where $\mathbf{a} = (2,0)$ and $\mathbf{b} = (1, \sqrt{3})$. Find the vectors $r[\pi/3](\mathbf{a})$, $r[\pi/3](\mathbf{b})$, $q[\pi/3](\mathbf{a})$, $q[\pi/3](\mathbf{b})$, $q[\pi/6](\mathbf{a})$ and $q[\pi/6](\mathbf{b})$, expressing your answers as linear combinations of $\mathbf{a}$ and $\mathbf{b}$. Then calculate the composite $f_1 f_2$ in the following cases, expressing your result in standard form.

(a) $f_1 = r[\pi/3]$, $f_2 = t[\mathbf{a}]\, r[\pi/3]$
(b) $f_1 = q[\pi/3]$, $f_2 = t[\mathbf{a} + \mathbf{b}]\, q[2\pi/3]$
(c) $f_1 = t[\mathbf{a}]\, r[\pi/3]$, $f_2 = t[\mathbf{b}]\, q[0]$
(d) $f_1 = t[\mathbf{a}]\, q[\pi/6]$, $f_2 = t[\mathbf{b}]$

When two symmetries of a lattice are specified in standard form, we now know how to form their composites. It is often the case, however, that a symmetry is described as being a rotation about a certain point in the plane or as being a reflection or glide reflection in a certain line in the plane. In order to write these symmetries in standard form, we need the formulae given by Equations 7–15 of the Isometry Toolkit. In each case, we shall give one worked example followed by two exercises.

Example 2.2

Consider the lattice $L = L(\mathbf{a}, \mathbf{b})$, where $\mathbf{a} = (2, -1)$ and $\mathbf{b} = (2, 2)$. Find the standard form of $r[\mathbf{c}, \pi]$ where $\mathbf{c} = \frac{1}{2}\mathbf{a} + \frac{1}{2}\mathbf{b}$, and show that this is a rotation of L.

Solution

By Equation 9 of the Isometry Toolkit,

$$r[\mathbf{c}, \pi] = t[2\mathbf{c}]\, r[\pi]$$
$$= t[\mathbf{a} + \mathbf{b}]\, r[\pi]$$
$$= t[(4,1)]\, r[\pi].$$

Now we saw in Subsection 2.1 that $r[\pi]$ is a symmetry of any lattice. Since $t[\mathbf{a} + \mathbf{b}] = t[\mathbf{a}]\, t[\mathbf{b}]$, it follows that $t[\mathbf{a} + \mathbf{b}]$ is a translation of L. Hence the composite, $t[\mathbf{a} + \mathbf{b}]\, r[\pi] = r[\mathbf{c}, \pi]$, is a rotation of L. ♦

Exercise 2.5

For each of the following rotations $r[\mathbf{c}, \theta]$ of a lattice $L(\mathbf{a}, \mathbf{b})$, where $\mathbf{a} = (1, 0)$ and $\mathbf{b} = (0, 1)$, write $r[\mathbf{c}, \theta]$ in standard form.

(a) $\mathbf{c} = (\mathbf{a} + \mathbf{b})/2$, $\theta = \pi/2$

(b) $\mathbf{c} = \mathbf{a} + \mathbf{b}$, $\theta = 3\pi/2$

Exercise 2.6

For each of the following rotations $r[\mathbf{c}, \theta]$ of a lattice $L(\mathbf{a}, \mathbf{b})$, where $\mathbf{a} = (2, 0)$ and $\mathbf{b} = (1, \sqrt{3})$, write $r[\mathbf{c}, \theta]$ in standard form.

(a) $\mathbf{c} = \mathbf{b}$, $\theta = \pi/3$

(b) $\mathbf{c} = (\mathbf{a} + \mathbf{b})/3$, $\theta = 2\pi/3$

Example 2.3

Consider the lattice $L = L(\mathbf{a}, \mathbf{b})$ where $\mathbf{a} = (2, -1)$ and $\mathbf{b} = (2, 1)$. Find the standard form of $q[\mathbf{c}, \pi/2]$ where $\mathbf{c} = \frac{1}{2}\mathbf{a} + \frac{1}{2}\mathbf{b}$, and show that this is a reflection of L.

Solution

By Equation 11 of the Isometry Toolkit,

$$q[\mathbf{c}, \pi/2] = t[\mathbf{d}]\, q[\pi/2],$$

where

$$\mathbf{d} = \mathbf{c} - q[\pi/2](\mathbf{c})$$
$$= (2, 0) - q[\pi/2](2, 0)$$
$$= (2, 0) - (-2, 0)$$
$$= (4, 0).$$

Thus,

$$q[\mathbf{c}, \pi/2] = t[(4, 0)]\, q[\pi/2].$$

Now,

$$q[\pi/2](\mathbf{a}) = (-2, -1) = -\mathbf{b}$$

and

$$q[\pi/2](\mathbf{b}) = (-2, 1) = -\mathbf{a},$$

and so $q[\pi/2]$ is a reflection of L.

Also,

$$\mathbf{d} = (4, 0) = \mathbf{a} + \mathbf{b},$$

and so $t[\mathbf{d}]$ is a translation of L.

Thus the composite, $q[\mathbf{c}, \pi/2]$, is a reflection of L. ♦

Exercise 2.7

For each of the following reflections $q[\mathbf{c}, \theta]$ of a lattice $L(\mathbf{a}, \mathbf{b})$, where $\mathbf{a} = (1, 0)$ and $\mathbf{b} = (0, 1)$, write $q[\mathbf{c}, \theta]$ in standard form.

(a) $\mathbf{c} = (\mathbf{a} + \mathbf{b})/2, \theta = \pi/2$
(b) $\mathbf{c} = \mathbf{a} + \mathbf{b}, \theta = 3\pi/4$

Exercise 2.8

For each of the following reflections $q[\mathbf{c}, \theta]$ of a lattice $L(\mathbf{a}, \mathbf{b})$, where $\mathbf{a} = (2, 0)$ and $\mathbf{b} = (1, \sqrt{3})$, write $q[\mathbf{c}, \theta]$ in standard form.

(a) $\mathbf{c} = \mathbf{b}, \theta = \pi/6$
(b) $\mathbf{c} = (\mathbf{a} + \mathbf{b})/3, \theta = \pi/2$

Example 2.4

Consider again the lattice $L = L(\mathbf{a}, \mathbf{b})$ where $\mathbf{a} = (2, -1)$ and $\mathbf{b} = (2, 1)$, and let f be the glide reflection

$$f = q[\mathbf{g}, \mathbf{c}, 0]$$

where $\mathbf{g} = \frac{1}{2}\mathbf{a} + \frac{1}{2}\mathbf{b}$ and $\mathbf{c} = \frac{1}{2}\mathbf{b}$.

State whether f is essential or inessential, and write it in standard form.

Solution

The translation component $t[\mathbf{g}] = t[\frac{1}{2}\mathbf{a} + \frac{1}{2}\mathbf{b}]$ is not a translation of L, so f is essential.

We proceed thus:

$$f = q[\tfrac{1}{2}\mathbf{a} + \tfrac{1}{2}\mathbf{b}, \tfrac{1}{2}\mathbf{b}, 0]$$
$$= t[\mathbf{d}]\, q[0] \quad \text{where } \mathbf{d} = \tfrac{1}{2}\mathbf{a} + \tfrac{1}{2}\mathbf{b} + \tfrac{1}{2}\mathbf{b} - \tfrac{1}{2}\mathbf{a}$$
$$= t[\mathbf{b}]\, q[0],$$

and this is in standard form.

Using Equation 14 of the Isometry Toolkit together with the fact that $q[0]\!\left(\tfrac{1}{2}\mathbf{b}\right) = \tfrac{1}{2}\mathbf{a}$.

♦

Exercise 2.9

For each of the following glide reflections $q[\mathbf{g}, \mathbf{c}, 0]$ of a lattice $L(\mathbf{a}, \mathbf{b})$, where $\mathbf{a} = (1, 0)$ and $\mathbf{b} = (0, 1)$, state whether it is essential or inessential and write it in standard form.

(a) $\mathbf{c} = \mathbf{b}, \theta = 0, \mathbf{g} = \mathbf{a}$.
(b) $\mathbf{c} = \tfrac{1}{2}\mathbf{b}, \theta = \pi/4, \mathbf{g} = \tfrac{1}{2}\mathbf{a} + \tfrac{1}{2}\mathbf{b}$.

Exercise 2.10

For each of the following glide reflections $q[\mathbf{g}, \mathbf{c}, 0]$ of a lattice $L(\mathbf{a}, \mathbf{b})$, where $\mathbf{a} = (2, 0)$ and $\mathbf{b} = (1, \sqrt{3})$, state whether it is essential or inessential and write it in standard form.

(a) $\mathbf{c} = \mathbf{a}, \theta = \pi/3, \mathbf{g} = \mathbf{b}$
(b) $\mathbf{c} = \tfrac{1}{2}\mathbf{b}, \theta = 0, \mathbf{g} = \tfrac{1}{2}\mathbf{a}$

As we have seen, writing the symmetries of a lattice in standard form is certainly very convenient when we are dealing with composites of symmetries. There is still the problem of identifying the geometrical features of a symmetry given in standard form. Although it may be clear that a given symmetry is a rotation through an angle θ, we would like to be able to find its centre. If it is a reflection, we would like to know its axis, and, if it is a glide reflection, we would like to know both its axis and its translation component along the axis. We shall see how this may be done algebraically when we study particular lattices in Section 3.

2.4 The crystallographic restriction

In this section, we prove an important theorem which shows that a plane lattice can have rotations of only certain orders.

We know already that every lattice possesses rotations of order 2. The lattice of integer points, which we looked at earlier, has rotations of orders 2 and 4. In the Introduction to this unit, you saw a wallpaper pattern which has a lot of different symmetries. It is based on a lattice which has rotations of orders 2, 3 and 6. The theorem that we prove can be generalized to a theorem about lattices in higher dimensions. This will not concern us in this unit, but the fact that there is a three-dimensional analogue will help to explain its name; it is of great importance to crystallographers.

> *Theorem 2.3 The crystallographic restriction*
>
> The order of a rotation of a plane lattice is equal to 1, 2, 3, 4 or 6.

Proof

Let L be a lattice which admits a rotation of order n. We know from Subsection 2.1 that if $n = 1$ or 2, then L does admit such a rotation. To prove the theorem, then, it will be sufficient to take $n \geq 3$ and show that the only possibilities for n are 3, 4 and 6. From our discussion in Subsection 2.2, we know that the lattice must have a rotation r whose angle is $2\pi/n$. Now choose a lattice point A whose vector $\mathbf{a}$ has least magnitude. The conjugate $r\,t[\mathbf{a}]\,r^{-1}$ of $t[\mathbf{a}]$ by the rotation r must be a translation of the lattice L. Let B be the point whose corresponding vector is $\mathbf{b} = r\,t[\mathbf{a}]\,r^{-1}(\mathbf{0}) = r(\mathbf{a})$. The line segments OA and OB will have the same length, and the angle between OA and OB will be $2\pi/n$. Let us examine the pair of vectors $\{\mathbf{a}, \mathbf{b}\}$ and show that they satisfy the minimality conditions.

By our choice of $\mathbf{a}$ the first minimality condition is satisfied and, since we have taken $n \geq 3$, the vectors $\mathbf{a}$ and $\mathbf{b}$ are linearly independent. Now $\|\mathbf{b}\| = \|\mathbf{a}\|$, so the second minimality condition is also satisfied. From Theorem 1.6(c) we get

$$\pi/3 \leq 2\pi/n \leq 2\pi/3$$

and this shows that $n \leq 6$.

So far, we have narrowed the possible orders of a rotation to the numbers 1, 2, 3, 4, 5 and 6. It remains to show that the case $n = 5$ cannot occur.

As in the argument above, we choose a lattice point A whose vector $\mathbf{a}$ has least magnitude — but, this time, we take the translation $r^2\,t[\mathbf{a}]\,r^{-2}$ and choose B to be the point represented by $\mathbf{b} = r^2\,t[\mathbf{a}]\,r^{-2}(\mathbf{0}) = r^2(\mathbf{a})$. The angle of the rotation r is $2\pi/5$, so the angle of r^2 is $4\pi/5$. Again $\{\mathbf{a}, \mathbf{b}\}$ must satisfy the minimality conditions so, by Theorem 1.6(c), we must have

$$\pi/3 \leq 4\pi/5 \leq 2\pi/3.$$

But this is clearly false. We have shown that the case $n = 5$ can never occur. This completes the proof. ∎

We shall find Theorem 2.3 extremely important when we come to study the classification of lattices and wallpaper patterns. When we are looking at a particular lattice L, we now know that there can only be rotations of L which have orders 1, 2, 3, 4 or 6. Can a given lattice have rotations with all these orders or are only certain combinations possible? The following theorem answers this question.

> **Theorem 2.4**
>
> For any given lattice L, the orders of the rotations of L form one of the following sets:
>
> (a) $\{1, 2\}$;
>
> (b) $\{1, 2, 4\}$;
>
> (c) $\{1, 2, 3, 6\}$.

Proof

We know that every lattice L will have rotations of orders 1 and 2, so these numbers must be included in each combination. Let us examine various cases.

Case 1   There are no rotations of orders 3, 4 or 6.

In this case, the orders which occur form the set $\{1, 2\}$.

Case 2   There is a rotation of order 4.

A rotation of order 4 has angle $\pm\pi/2$, so if L has a rotation of this order, then it or its inverse is a rotation r with angle $-\pi/2$.

If L also possessed a rotation of order 3, then it would have angle $\pm 2\pi/3$, so it or its inverse would be a rotation s with angle $2\pi/3$. Then rs would be a rotation of L with angle $\frac{2\pi}{3} - \frac{\pi}{2} = \frac{\pi}{6}$. This is of order 12, contradicting Theorem 2.3. Thus L cannot possess a rotation of order 3.

If L possessed a rotation of order 6, then its square would be of order 3, and we have just ruled out this possibility.

Thus if L has a rotation of order 4, the rotations which occur are exactly the set $\{1, 2, 4\}$.

Case 3   There is a rotation of order 3.

A rotation of order 3 has angle $\pm 2\pi/3$, so if L has a rotation of this order, then it or its inverse is a rotation s with angle $-2\pi/3$.

Now, from Subsection 2.1, L must have a rotation r of order 2 and therefore angle π. Thus, rs is a rotation of L with angle $\pi - \frac{2\pi}{3} = \frac{\pi}{3}$. The order of rs is 6, so we have shown that if L possesses rotations of order 3, it must also have rotations of order 6. Order 4 is ruled out by *Case 2*, so the orders which occur must be exactly the set $\{1, 2, 3, 6\}$.

Case 4   There is a rotation of order 6.

Since the square of such a rotation will have order 3, this case has already been dealt with above. The orders which occur are $\{1, 2, 3, 6\}$.

We have examined all possibilities, so our proof is complete.   ∎

You have now had some practice at handling the symmetries of a lattice. This experience will be needed in the next section, where we examine particular types of lattices and identify five different geometric types. Some lattices have a great deal of symmetry while others have only the minimum. We start with the least complicated of them.

3 FIVE TYPES OF PLANE LATTICES

3.1 The parallelogram lattice

A **parallelogram lattice** is a type of lattice whose symmetries consist solely of translations and rotations of order 2. It has no rotations of order greater than 2 and it admits no reflections or glide reflections.

We shall see shortly that this implies that if $\{\mathbf{a}, \mathbf{b}\}$ is a basis, then $\|\mathbf{a}\| \neq \|\mathbf{b}\|$, and $\mathbf{a}$ and $\mathbf{b}$ are not orthogonal.

An example of a parallelogram lattice is $L(\mathbf{a}, \mathbf{b})$, where $\mathbf{a} = (1, 2)$ and $\mathbf{b} = (-2, 2)$. This is illustrated in Figure 3.1.

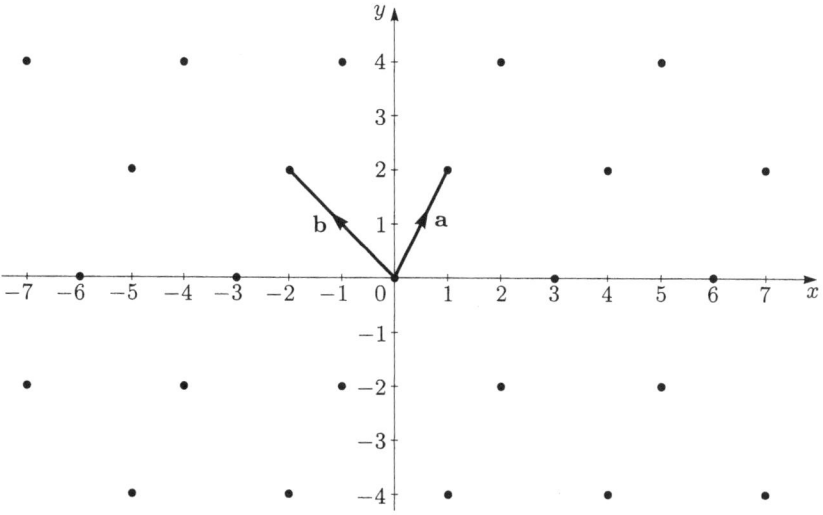

Figure 3.1

In Subsection 2.1, we showed that every lattice has rotations of order 2. They can all be written in standard form as $t\, r[\pi]$, where t is some translation of the lattice. If we are given a rotation of order 2 in standard form, it is easy to find its centre. Try the following exercise, which revises the techniques of *Unit IB1*, Section 5.

Exercise 3.1

Find the centre of the rotation $t[\mathbf{d}]\, r[\pi]$ when:
(a) $t[\mathbf{d}] = t[\mathbf{a}]$;
(b) $t[\mathbf{d}] = t[\mathbf{a}]\, t[\mathbf{b}]$;
(c) $t[\mathbf{d}] = (t[\mathbf{a}])^n\, (t[\mathbf{b}])^m$.

The centres of rotations of order 2 are called **2-centres**. From the above exercise, they all have the form $\frac{1}{2}n\mathbf{a} + \frac{1}{2}m\mathbf{b}$, where n and m are integers.

Theorem 3.1

For every lattice $L(\mathbf{a}, \mathbf{b})$, the set of 2-centres is the lattice $L\left(\frac{1}{2}\mathbf{a}, \frac{1}{2}\mathbf{b}\right)$.

In a basic parallelogram, a 2-centre is a vertex or the centre of the parallelogram or the midpoint of an edge. For any parallelogram which is a translation of a basic parallelogram, the same remarks apply.

In later parts of this section, we shall be looking at points which are centres for rotations with orders 3, 4 and 6. These points will be called **3-centres**, **4-centres** and **6-centres**, respectively. Note that we do not exclude the possibility that a point in $\mathbb{R}^2$ may be, for example, both a 2-centre and a 4-centre. In fact, a 4-centre will always be a 2-centre, since the square of a rotation of order 4 is a rotation of order 2 with the same centre.

As the parallelogram lattice has so few symmetries, there is little more to say about it. It does occur quite frequently as the underlying lattice of a wallpaper pattern. In Figure 3.2, we show an example. The symmetries of the pattern are the same as those of the underlying lattice (i.e. the set of square dots embedded in the lattice).

Figure 3.2

You should examine this figure and check that it has 2-centres at the stated points.

Clearly, every plane lattice must include among its symmetries those of a parallelogram lattice. We now look at lattices which, although possessing no rotations of order higher than 2, do admit some reflections and glide reflections. It turns out that there are exactly two types. This is a consequence of a theorem which we prove in Subsection 5.3.

3.2 The rectangular lattice

A **rectangular lattice** has a basis consisting of vectors **a** and **b** which are orthogonal. Such a basis is said to be an **orthogonal basis**. We also require the magnitudes $\|\mathbf{a}\|$ and $\|\mathbf{b}\|$ to be different. The corresponding basic parallelogram will be a rectangle. In Figure 3.3, we show an example of such a lattice where $\mathbf{a} = (1, 0)$ and $\mathbf{b} = (0, 2)$.

If $\|\mathbf{a}\| = \|\mathbf{b}\|$, then (as we shall see in Subsection 3.4) there are rotational symmetries of order 4, and the lattice is a square lattice.

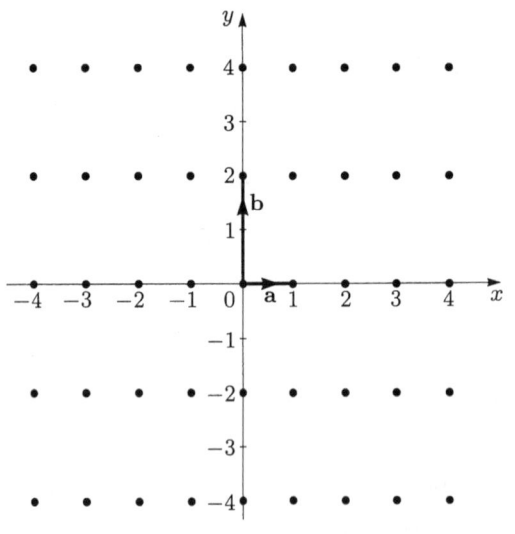

Figure 3.3

The 2-centres are located at the points $\frac{1}{2}n\mathbf{a} + \frac{1}{2}m\mathbf{b}$, where n and m are integers. In our example, these are the points $\left(\frac{1}{2}n, m\right)$. You may care to mark some of these on the figure.

By studying this lattice, you will find that there are reflections in lines through each 2-centre of the lattice and that these axes are parallel to the x-axis or parallel to the y-axis.

There are glide reflections as well. We shall now show that, in the case of the rectangular lattice, they are all inessential glide reflections.

Theorem 3.2

Let L be a rectangular lattice. Then every glide reflection of L is inessential.

Proof

Let $\{\mathbf{a}, \mathbf{b}\}$ be a reduced basis, and choose the axes so that $\mathbf{a}$ and $\mathbf{b}$ are in the directions of the x-and y-axes. Then $q[0]$ and $q[\pi/2]$ are the only reflection symmetries of the lattice which fix $\mathbf{0}$. (To see this, note that if $q[\theta]$ is a symmetry, then so is $q[\theta]\,q[0]$, which by Equation 3 of the Isometry Toolkit is the rotation $r[2\theta]$. But the only rotation symmetries of L are of order 1 or 2.)

Thus, by Theorem 2.2, every glide reflection symmetry of L is of the form $t[\mathbf{d}]\,q[0]$ or $t[\mathbf{d}]\,q[\pi/2]$, where $\mathbf{d}$ is some integer combination

$$\mathbf{d} = n\mathbf{a} + m\mathbf{b}.$$

Now

$$t[\mathbf{d}]\,q[0] = t[n\mathbf{a}]\,(t[m\mathbf{b}]\,q[0])$$
$$= t[n\mathbf{a}]\,q\left[\tfrac{1}{2}m\mathbf{b}, 0\right],$$

and hence the translation component of $t[\mathbf{d}]\,q[0]$ is $t[n\mathbf{a}]$, which is a symmetry of L, and $t[\mathbf{d}]\,q[0]$ is an inessential glide reflection.

Similarly,

$$t[\mathbf{d}]\,q[\pi/2] = t[m\mathbf{b}]\,(t[n\mathbf{a}]\,q[\pi/2])$$
$$= t[m\mathbf{b}]\,q\left[\tfrac{1}{2}n\mathbf{a}, \pi/2\right],$$

and again the translation component, $t[m\mathbf{b}]$, is a symmetry of L, so $t[\mathbf{d}]\,q[\pi/2]$ is an inessential glide reflection. ∎

In the following exercise, we ask you to obtain expressions for some of the reflections of the lattice.

Exercise 3.2

Let L be a rectangular lattice with an orthogonal basis $\{\mathbf{a}, \mathbf{b}\}$. Let A and B be the points of $\mathbb{R}^2$ corresponding to $\mathbf{a}$ and $\mathbf{b}$, and let q be the reflection in the line through O and A. Find expressions for the reflections q_1, q_2, q', q'_1 and q'_2, shown in Figure 3.4, in terms of q, $r[\pi]$, $t[\mathbf{a}]$ and $t[\mathbf{b}]$.

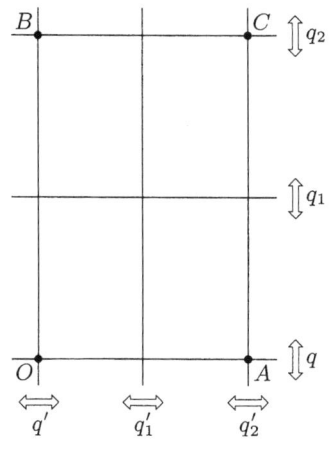

Figure 3.4

The next type of lattice that we shall investigate is the *rhombic lattice*.

3.3 The rhombic lattice

The other type of lattice which possesses no rotation with order greater than 2 is the **rhombic lattice**. This lattice has a basis $\{\mathbf{a}, \mathbf{b}\}$ consisting of vectors $\mathbf{a}$ and $\mathbf{b}$ where $\|\mathbf{a}\| = \|\mathbf{b}\|$ and where the angle between $\mathbf{a}$ and $\mathbf{b}$ is not $\pi/3$, $\pi/2$ or $2\pi/3$. This restriction is made in order to exclude lattices which have rotations of orders 3, 4 or 6. Notice that the pair $\{\mathbf{a}, \mathbf{b}\}$ is not necessarily a reduced basis for the lattice. It may well happen that one of the vectors $\mathbf{a} + \mathbf{b}$ or $\mathbf{a} - \mathbf{b}$ has magnitude less than the magnitudes of $\mathbf{a}$ and $\mathbf{b}$. An example of a rhombic lattice is the lattice $L(\mathbf{a}, \mathbf{b})$, where $\mathbf{a} = (2, -1)$ and $\mathbf{b} = (2, 1)$ (see Figure 3.5).

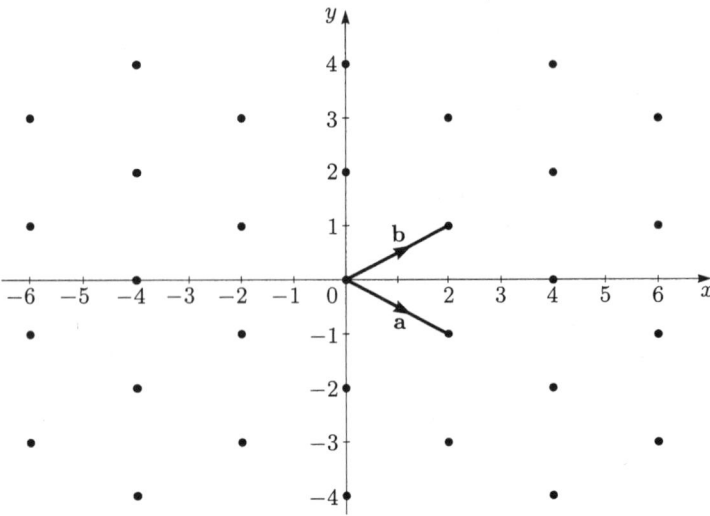

Figure 3.5

We have considered this lattice already. It is the lattice we used in Examples 2.1, 2.3 and 2.4. The 2-centres are located at the points $\frac{1}{2}n\mathbf{a} + \frac{1}{2}m\mathbf{b}$ and these are the points $(n + m, (m - n)/2)$. The reflections of the lattice which fix O are $q[0]$ and $q[\pi/2]$ (as with the rectangular lattice), so all the indirect symmetries can be expressed either as $t\,q[0]$ or as $t\,q[\pi/2]$, where t is a translation of the lattice. In Figure 3.6 we show the lattice again with the axes of reflection drawn as solid lines. In addition, there are glide reflections of the lattice whose axes lie midway between these lines. These are shown as broken lines in the figure.

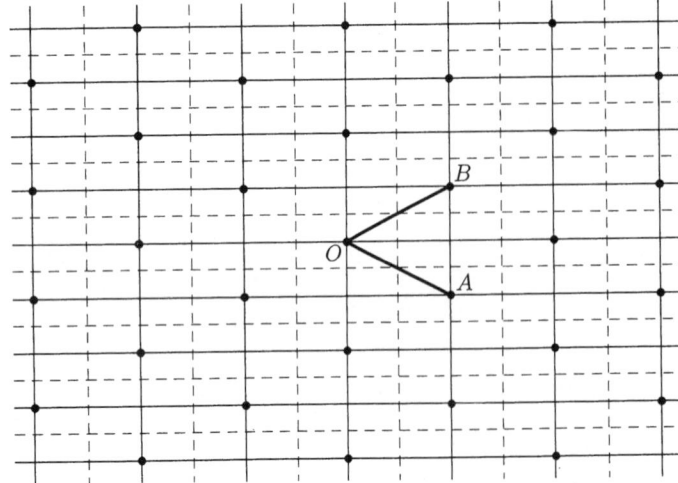

Figure 3.6

Notice that the reflection axes all pass through lattice points, while the glide axes contain no lattice points.

There is another name for a rhombic lattice. It is sometimes called a **centred rectangular lattice**. This is because we can obtain a rhombic lattice from a rectangular lattice by adding extra lattice points at the centre of each rectangle. If you look at Figure 3.6, you should be able to find rectangles whose vertices are lattice points and whose centres are also lattice points.

Incidentally, we can similarly construct a rectangular lattice from a rhombic lattice by adding extra lattice points at the centre of each rhombus.

From now on, we shall use the term **reflection axes** for axes that are axes of reflection symmetries, and **glide axes** for axes that are axes of glide reflection symmetries but are not axes of any reflection symmetries.

Exercise 3.3

Let L be a rhombic lattice with a basis $\{\mathbf{a}, \mathbf{b}\}$ such that $\|\mathbf{a}\| = \|\mathbf{b}\|$. Let q be the reflection in the line through O containing the point $\mathbf{a} + \mathbf{b}$ and let q' be the reflection in the line through O containing the point $\mathbf{a} - \mathbf{b}$. Figure 3.7 shows the reflections q and q', the axes of two other reflections q_1 and q_1' and the axes of two glide reflections g and g' both of which map O to B. Express the symmetries q_1, q', q_1', g and g' in terms of $t[\mathbf{a}]$, $t[\mathbf{b}]$, q and $r[\pi]$.

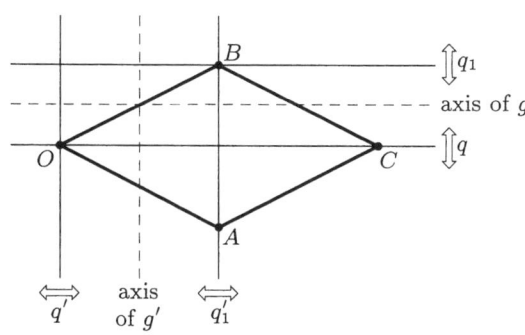

Figure 3.7

The next type of lattice that we shall investigate is the *square lattice*.

3.4 The square lattice

A **square lattice** L possesses rotations of order 4. We now show that L has a reduced basis $\{\mathbf{a}, \mathbf{b}\}$ where $\|\mathbf{a}\| = \|\mathbf{b}\|$ and now $\mathbf{a}$ and $\mathbf{b}$ are orthogonal.

Let $\mathbf{a}$ be a vector of L with least magnitude, and choose the x-axis in the direction of $\mathbf{a}$. If L has a rotation of order 4, then the rotation $r[\pi/2]$ will be a symmetry of L. Let $\mathbf{b} = r[\pi/2](\mathbf{a})$. Then $\mathbf{b}$ will be orthogonal to $\mathbf{a}$ and we have $\|\mathbf{b}\| = \|\mathbf{a}\|$. We can see that $\{\mathbf{a}, \mathbf{b}\}$ is thus a reduced basis for L which is orthogonal. We have already seen an example of this type of lattice. The lattice of integer points is a square lattice which has a reduced basis consisting of the vectors $\mathbf{a} = (1, 0)$ and $\mathbf{b} = (0, 1)$.

When we choose our x-axis to lie in the direction of the vector $\mathbf{a}$, the rotations with centre O will be e, $r[\pi/2]$, $r[\pi]$ and $r[3\pi/2]$. As in all cases, the 2-centres are located at the points $\frac{1}{2}n\mathbf{a} + \frac{1}{2}m\mathbf{b}$ ($n, m \in \mathbb{Z}$).

The rotations of order 4 will be of the form $t\,r[\pi/2]$ or $t\,r[3\pi/2]$, where t is a translation of the lattice. Writing t as $t[\mathbf{d}]$, the centre $\mathbf{c}$ of $t[\mathbf{d}]\,r[\theta]$ is related to $\mathbf{d}$ by the equation

$$\mathbf{d} = \mathbf{c} - r[\theta](\mathbf{c}),$$

where $\theta = \pi/2$ or $3\pi/2$. This equation does not, however, give $\mathbf{c}$ explicitly in terms of $\mathbf{d}$. In order to do this, we must do some calculations.

See Equation 8 of the Isometry Toolkit.

Let us take the case where $\theta = \pi/2$. We find that $r[\pi/2](\mathbf{a}) = \mathbf{b}$ and $r[\pi/2](\mathbf{b}) = -\mathbf{a}$. The matrix $\mathbf{A}$ which represents the rotation $r[\pi/2]$ with respect to the basis $\{\mathbf{a}, \mathbf{b}\}$ is $\begin{bmatrix} 0 & -1 \\ 1 & 0 \end{bmatrix}$. Let us put $\mathbf{c} = c_1\mathbf{a} + c_2\mathbf{b}$ and $\mathbf{d} = d_1\mathbf{a} + d_2\mathbf{b}$. Then the equation $\mathbf{d} = \mathbf{c} - r[\pi/2](\mathbf{c})$ can be written in matrix form as

$$\begin{bmatrix} d_1 \\ d_2 \end{bmatrix} = (\mathbf{I} - \mathbf{A}) \begin{bmatrix} c_1 \\ c_2 \end{bmatrix} = \begin{bmatrix} 1 & 1 \\ -1 & 1 \end{bmatrix} \begin{bmatrix} c_1 \\ c_2 \end{bmatrix}.$$

It follows from this that

$$\begin{bmatrix} c_1 \\ c_2 \end{bmatrix} = (\mathbf{I} - \mathbf{A})^{-1} \begin{bmatrix} d_1 \\ d_2 \end{bmatrix}$$

$$= \begin{bmatrix} \frac{1}{2} & -\frac{1}{2} \\ \frac{1}{2} & \frac{1}{2} \end{bmatrix} \begin{bmatrix} d_1 \\ d_2 \end{bmatrix}.$$

For any given $\mathbf{d}$ we can use this matrix equation to find the rotation centre $\mathbf{c}$.

Exercise 3.4

Using the matrix equation above, find the centre of the rotation $t[\mathbf{d}] \, r[\pi/2]$ for each of the following vectors $\mathbf{d}$.

(a) $\mathbf{d} = \mathbf{a}$
(b) $\mathbf{d} = \mathbf{b}$
(c) $\mathbf{d} = \mathbf{a} + \mathbf{b}$

Let us write the expression for $\mathbf{c}$ in terms of $\mathbf{d}$ as $\mathbf{c} = f[\theta](\mathbf{d})$. Then $f[\theta]$ is a linear map so, with $\mathbf{d} = n\mathbf{a} + m\mathbf{b}$, we have

$$\mathbf{c} = f[\theta](n\mathbf{a} + m\mathbf{b})$$
$$= nf[\theta](\mathbf{a}) + mf[\theta](\mathbf{b}).$$

We have found $f[\theta](\mathbf{a})$ and $f[\theta](\mathbf{b})$ for the case $\theta = \pi/2$ (see the solution to Exercise 3.4), so it follows that

$$f[\pi/2](\mathbf{d}) = n\left(\tfrac{1}{2}\mathbf{a} + \tfrac{1}{2}\mathbf{b}\right) + m\left(-\tfrac{1}{2}\mathbf{a} + \tfrac{1}{2}\mathbf{b}\right)$$
$$= \tfrac{1}{2}(n - m)\mathbf{a} + \tfrac{1}{2}(n + m)\mathbf{b}.$$

Notice that the numbers $n - m$ and $n + m$ are either both even or both odd. In the case when they are both even, the 4-centre is a lattice point. In the case when they are both odd, the 4-centre is at the centre of a square of the lattice. It is easily seen that all such points are 4-centres.

Exercise 3.5

Find the expression for $\mathbf{c} = f[\theta](\mathbf{d})$ in terms of $\mathbf{d} = n\mathbf{a} + m\mathbf{b}$ when $\theta = 3\pi/2$.

We see from this exercise that, although the expressions are different, the centres of rotations for $\theta = 3\pi/2$ are the same as those for $\theta = \pi/2$. This is hardly surprising, since every 4-centre must have rotations through $\pi/2$ and $3\pi/2$. We state our findings as a theorem.

> **Theorem 3.3**
>
> The 4-centres of a square lattice $L(\mathbf{a}, \mathbf{b})$, where $\{\mathbf{a}, \mathbf{b}\}$ is an orthogonal basis, form the lattice $L\left(\tfrac{1}{2}(\mathbf{a} + \mathbf{b}), \tfrac{1}{2}(-\mathbf{a} + \mathbf{b})\right)$.

Let us now investigate the indirect symmetries of a square lattice. It seems reasonable to expect that a square lattice $L(\mathbf{a}, \mathbf{b})$ will have the reflections and glide reflections of both the rectangular lattice and the rhombic lattice. This is indeed true. We may assume that the x-axis is chosen in the direction of **a** and that **b** is chosen to be $r[\pi/2](\mathbf{a})$. Then the reflections in lines through O will be $q[0]$, $q[\pi/4]$, $q[\pi/2]$ and $q[3\pi/4]$. All the indirect symmetries of L will be of the form tq, where t is some translation of L and q is one of these reflections. Figure 3.8 shows the basis vectors $OA = \mathbf{a}$ and $OB = \mathbf{b}$ for the lattice, and the reflection and glide axes. Each of these axes is parallel to one of the vectors **a**, **b**, **a** + **b** or **a** − **b**.

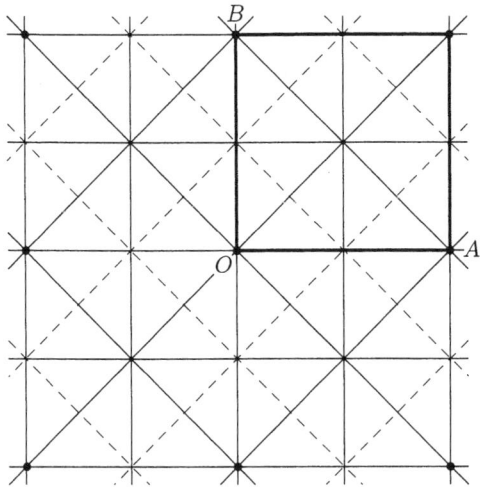

Figure 3.8

Using the formulae established in Subsection 2.3, it is possible to express each of the symmetries in standard form.

We shall now consider the last of the lattice types.

3.5 The hexagonal lattice

A **hexagonal lattice** has a reduced basis $\{\mathbf{a}, \mathbf{b}\}$ where $\|\mathbf{a}\| = \|\mathbf{b}\|$ and the angle between **a** and **b** is $\pi/3$. We can show that every lattice L which has a symmetry of order 3 will be of this type. Choose **a** to be a vector of L which has least magnitude. If there is a rotation of order 3, then we know from Theorem 2.4 that there is also a rotation of order 6, so the rotation $r[\pi/3]$ is a symmetry of L. If we choose $\mathbf{b} = r[\pi/3](\mathbf{a})$, then $\{\mathbf{a}, \mathbf{b}\}$ is a reduced basis for L with $\|\mathbf{b}\| = \|\mathbf{a}\|$. We have seen an example of a lattice of this type in Exercises 2.6, 2.8 and 2.10. The lattice $L(\mathbf{a}, \mathbf{b})$, where $\mathbf{a} = (2, 0)$ and $\mathbf{b} = (1, \sqrt{3})$, is a hexagonal lattice.

Figure 3.9 shows a hexagonal lattice with the basis vectors **a** and **b** chosen so that the angle between them is $\pi/3$.

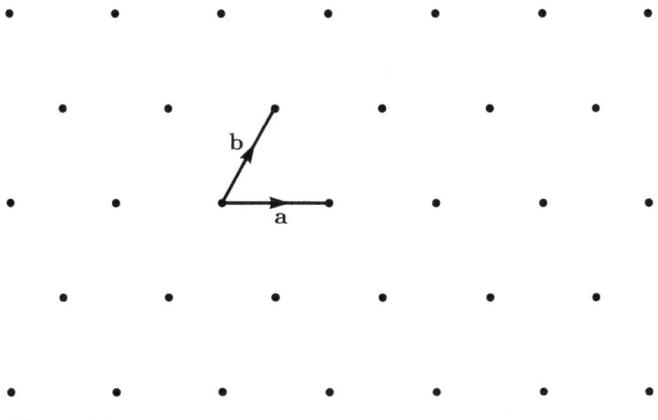

Figure 3.9

You should notice that we could, alternatively, choose a reduced basis $\{\mathbf{a}, \mathbf{b}\}$ where the angle between the vectors $\mathbf{a}$ and $\mathbf{b}$ is $2\pi/3$. For example, $\{(2,0),(-1,\sqrt{3})\}$ is a reduced basis for a hexagonal lattice.

We know that the 2-centres of any lattice are at the points $\frac{1}{2}n\mathbf{a} + \frac{1}{2}m\mathbf{b}$.

The rotations of order 6 will be of the form $t\,r[\pi/3]$ or $t\,r[5\pi/3]$ where t is a translation of the lattice. Since every 6-centre has a rotation through $\pi/3$ as well as a rotation through $5\pi/3$, we need only find the centres of of the rotations of form $t\,r[\pi/3]$; these will constitute all the 6-centres.

The centre $\mathbf{c}$ of $t[\mathbf{d}]\,r[\pi/3]$ is related to $\mathbf{d}$ by the equation

$$\mathbf{d} = \mathbf{c} - r[\pi/3](\mathbf{c})$$

and we must now express $\mathbf{c}$ in terms of $\mathbf{d}$.

Rather surprisingly, this expression turns out to be

$$\mathbf{c} = r[\pi/3](\mathbf{d}).$$

An easy way to see this is to observe that the vectors $\mathbf{0}$, $\mathbf{c}$ and $\mathbf{d}$ form the vertices of an equilateral triangle (see Figure 3.10).

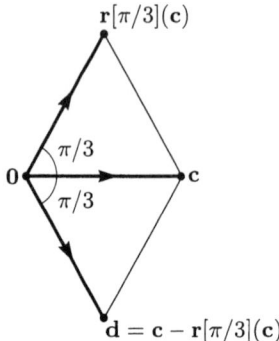

Figure 3.10

In the next exercise, we ask you to establish this result algebraically.

Exercise 3.6

Consider the usual matrix representation

$$\mathbf{A} = \begin{bmatrix} \frac{1}{2} & -\frac{\sqrt{3}}{2} \\ \frac{\sqrt{3}}{2} & \frac{1}{2} \end{bmatrix}$$

for $r[\pi/3]$. Show that

$$(\mathbf{I} - \mathbf{A})^{-1} = \mathbf{A},$$

and hence show that $\mathbf{c} = r[\pi/3](\mathbf{d})$.

In fact, the manipulation is easier if, instead of the usual orthogonal basis for $\mathbb{R}^2$, we actually take the vectors $\mathbf{a}$ and $\mathbf{b}$. Recalling that $r[\pi/3](\mathbf{a}) = \mathbf{b}$ and noting that $r[\pi/3](\mathbf{b}) = \mathbf{b} - \mathbf{a}$, we see that the matrix $\mathbf{A}'$ of $r[\pi/3]$ with respect to the basis $\{\mathbf{a}, \mathbf{b}\}$ of $\mathbb{R}^2$ is

$$\mathbf{A}' = \begin{bmatrix} 0 & -1 \\ 1 & 1 \end{bmatrix}.$$

Then

$$\mathbf{I} - \mathbf{A}' = \begin{bmatrix} 1 & 1 \\ -1 & 0 \end{bmatrix},$$

which is easily seen to equal $(\mathbf{A}')^{-1}$.

Thus, if we have a rotation $r[\mathbf{c}, \pi/3]$, expressed in standard form as $t[\mathbf{d}]\, r[\pi/3]$, and if $\mathbf{c}$ and $\mathbf{d}$ are expressed with respect to the basis $\{\mathbf{a}, \mathbf{b}\}$ as

$$\mathbf{c} = c_1 \mathbf{a} + c_2 \mathbf{b}, \quad \mathbf{d} = d_1 \mathbf{a} + d_2 \mathbf{b},$$

then we obtain the simple relationships

$$\begin{bmatrix} d_1 \\ d_2 \end{bmatrix} = \begin{bmatrix} 1 & 1 \\ -1 & 0 \end{bmatrix} \begin{bmatrix} c_1 \\ c_2 \end{bmatrix},$$

$$\begin{bmatrix} c_1 \\ c_2 \end{bmatrix} = \begin{bmatrix} 0 & -1 \\ 1 & 1 \end{bmatrix} \begin{bmatrix} d_1 \\ d_2 \end{bmatrix},$$

that is,

$$d_1 = c_1 + c_2, \quad d_2 = -c_1 \tag{3.1}$$

and

$$c_1 = -d_2, \quad c_2 = d_1 + d_2. \tag{3.2}$$

Now the rotational symmetries $r[\mathbf{c}, \pi/3]$ of L are those of the form $t[\mathbf{d}]\, r[\pi/3]$ where $\mathbf{d} = n\mathbf{a} + m\mathbf{b}$ is a translation of the lattice. Setting $d_1 = n$ and $d_2 = m$ in Equation 3.2 gives us

$$c_1 = -m, \quad c_2 = n + m,$$

so that

$$\mathbf{c} = -m\mathbf{a} + (n+m)\mathbf{b}.$$

These are the points of $\mathbb{R}^2$ which are 6-centres, and they are lattice points. From Equation 3.1, we see that for every integer choice of c_1 and c_2 there is a corresponding translation $\mathbf{d}$ of the lattice, so that all the lattice points are 6-centres. This establishes the following theorem.

> **Theorem 3.4**
>
> For a hexagonal lattice $L(\mathbf{a}, \mathbf{b})$, the set of 6-centres is the lattice $L(\mathbf{a}, \mathbf{b})$ itself.

We now look for the 3-centres of L. If $\mathbf{c}$ is a 3-centre then there is a rotation of L of the form $r[\mathbf{c}, 2\pi/3]$. In standard form, this is $t[\mathbf{d}]\, r[2\pi/3]$, where $\mathbf{d}$ is some lattice point.

Since $r[2\pi/3](\mathbf{a}) = \mathbf{b} - \mathbf{a}$ and $r[2\pi/3](\mathbf{b}) = -\mathbf{a}$, the matrix for $r[2\pi/3]$ with respect to $\{\mathbf{a}, \mathbf{b}\}$ is

$$\mathbf{A} = \begin{bmatrix} -1 & -1 \\ 1 & 0 \end{bmatrix}$$

so the matrix $\mathbf{I} - \mathbf{A}$ is $\begin{bmatrix} 2 & 1 \\ -1 & 1 \end{bmatrix}$.

Hence

$$(\mathbf{I} - \mathbf{A})^{-1} = \begin{bmatrix} 1/3 & -1/3 \\ 1/3 & 2/3 \end{bmatrix}.$$

Taking an arbitrary lattice point $\mathbf{d} = n\mathbf{a} + m\mathbf{b}$, we find that the centre of the rotation $t[\mathbf{d}]\, r[2\pi/3]$ lies at the point

$$\mathbf{c} = f[2\pi/3](\mathbf{d}) = n(\mathbf{a} + \mathbf{b})/3 + m(2\mathbf{b} - \mathbf{a})/3.$$

> **Theorem 3.5**
>
> For a hexagonal lattice, the set of 3-centres is the lattice $L((\mathbf{a} + \mathbf{b})/3, (2\mathbf{b} - \mathbf{a})/3)$.

Note that the lattice points of a hexagonal lattice are 6-centres, 3-centres and 2-centres, according to our definitions.

Exercise 3.7

Find the 3-centres which lie on or inside the basic parallelogram corresponding to the basis $\{\mathbf{a}, \mathbf{b}\}$, where the angle between $\mathbf{a}$ and $\mathbf{b}$ is $\pi/3$. Write in standard form all these rotations of order 3 at the points which are not 6-centres.

Let us now examine the reflections and glide reflections of a hexagonal lattice. We may as well take the lattice $L = L(\mathbf{a}, \mathbf{b})$, where $\mathbf{a} = (2, 0)$ and $\mathbf{b} = (1, \sqrt{3})$. There are six reflections of L which fix the origin O. These are $q[0]$, $q[\pi/6]$, $q[\pi/3]$, $q[\pi/2]$, $q[2\pi/3]$ and $q[5\pi/6]$. All the indirect symmetries of L are of the form tq, where t is a translation of L and q is one of these six reflections. Through each lattice point there are six reflections parallel to the six reflections through O. These are shown as solid lines in Figure 3.11.

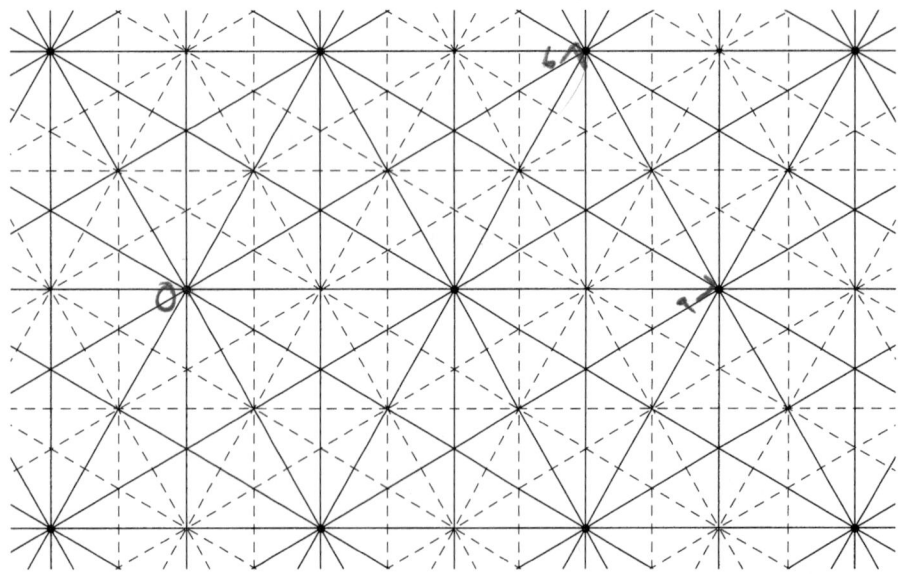

Figure 3.11

All the reflection axes will also be axes for inessential glide reflections obtained by composing a reflection with a translation along its axis. In addition, there are glide axes shown by broken lines. An example of a glide reflection corresponding to one of these is $t[\mathbf{b}]\,q[0]$. Its axis passes through the point $\mathbf{b}/2$ and is parallel to the vector $\mathbf{a}$.

It is worth looking at the points of $\mathbb{R}^2$ which are intersections of axes: there are three types of point. The 6-centres have six reflection axes. The 3-centres which are not 6-centres have three reflection axes. Finally, the 2-centres which are not 6-centres have two reflection axes and four glide axes.

We have now discussed five types of lattice that have distinct symmetry properties. We have not yet proved that there are these types and no others in two dimensions. The proof is fairly involved, and (like the corresponding proof concerning frieze types in *Unit IB3*) is best presented at the end of the unit, after you have had a chance to familiarize yourself with the properties of all five lattice types.

The next section is concerned with obtaining a precise and useful description of $\Gamma(L)$, in terms of generators and relations, for each of the five plane lattice types. Then, in Section 5, we prove that there are no further plane lattice types.

4 THE PLANE LATTICE GROUPS

In this section, we consider each of the five lattices and give a description of its symmetry group in terms of generators and relations. You may recall, from *Unit IB2*, Section 3, that a set S of elements of a group G will generate G if every element of G can be expressed as a composite involving elements of S and their inverses. To see whether or not two such composites represent the same element, it is useful to write elements in a standard form. This is where the relations of the group are needed.

We shall also be looking at various group actions, using the ideas of orbits and stabilizers which you met in *Units IB2* and *GE1*.

If L is any lattice, we recall that its full symmetry group is denoted by $\Gamma(L)$, its direct symmetry group by $\Gamma^+(L)$, and its translation group by $\Delta(L)$.

4.1 The parallelogram lattice

The only symmetries of a parallelogram lattice L are the translations and the rotations of order 2. The translations constitute the group $\Delta(L)$, which is generated by the pair $\{t[\mathbf{a}], t[\mathbf{b}]\}$, where $\{\mathbf{a}, \mathbf{b}\}$ is any basis for L. As $\Delta(L)$ is generated by two translations, we shall denote it by T_2:

$$T_2 = \{(t[\mathbf{a}])^n (t[\mathbf{b}])^m : n, m \in \mathbb{Z}\}.$$

In *Unit IB3* we shortened the notation $t[\mathbf{a}]$ to just t. In this two-dimensional context, we shall use t_a and t_b:

$$t[\mathbf{a}] = t_a, \quad t[\mathbf{b}] = t_b.$$

Now, by Equation 1 of the Isometry Toolkit, t_a and t_b commute. Adding this information to our description of T_2, we have

$$T_2 = \{t_a^n t_b^m : n, m \in \mathbb{Z};\ t_a t_b = t_b t_a\}.$$

In Unit IB3, we denoted the translation group of a frieze (which is generated by one translation) by T_1.

Exercise 4.1

Write down an isomorphism from T_2 to the group $\mathbb{Z} \times \mathbb{Z}$.

Each rotation of order 2 of L can be written as $t\, r[\pi]$, where t belongs to T_2. As there are no other symmetries of the parallelogram lattice, the symmetry group $\Gamma(L)$ is the group generated by the set

$$\{t_a, t_b, r\}$$

where we have also shortened the notation $r[\pi]$ to r.

This follows from Theorem 2.2, since t is the translation part and $r[\pi]$ the linear part.

The standard form which we shall adopt for symmetries of the parallelogram lattice is $t_a^n t_b^m$ for a translation and $t_a^n t_b^m r$ for a rotation. To express an arbitrary symmetry in this form, we need some additional relations, namely:

$$rt_a = t_a^{-1} r;$$
$$rt_b = t_b^{-1} r;$$
$$r^2 = e.$$

From the first two of these we deduce that

$$r\, t[\mathbf{d}] = (t[\mathbf{d}])^{-1} r$$

for all translations $t[\mathbf{d}]$. We may then write a composite of rotations, $(t[\mathbf{c}]\, r)(t[\mathbf{d}]\, r)$, as

$$(t[\mathbf{c}]\, r)(t[\mathbf{d}]\, r) = t[\mathbf{c}] (t[\mathbf{d}])^{-1} r^2$$
$$= t[\mathbf{c}] (t[\mathbf{d}])^{-1} \quad \text{(since } r^2 = e\text{)}.$$

This result can, of course, be deduced from Equation 6 of the Isometry Toolkit.

Thus, we may describe $\Gamma(L)$ formally, in terms of generators and relations, as follows:

$$\Gamma(L) = \langle t_a, t_b, r : t_a t_b = t_b t_a,\ rt_a = t_a^{-1} r,\ rt_b = t_b^{-1} r,\ r^2 = e \rangle.$$

Recalling that the standard form for an element of $\Gamma(L)$ is $t_a^n t_b^m$ or $t_a^n t_b^m r$, there is a simpler way of writing $\Gamma(L)$, analogous to the way we wrote $\Gamma(F_i)$ for each of the frieze types F_i $(i = 1, \ldots, 7)$. It is

$$\Gamma(L) = \{xy : x \in T_2,\ y \in C_2;\ rx = x^{-1} r\},$$

where $C_2 = \{e, r\}$, the group of rotations of L with centre O.

In this form, we assume that the generators and relations for the groups T_2 and C_2 are known. The relation stated above connects an element from each of these groups, and means that $rx = x^{-1} r$ for *any* $x \in T_2$. It could, alternatively, be written as the pair of relations

$$rt_a = t_a^{-1} r,\quad rt_b = t_b^{-1} r.$$

In Subsection 3.3, you gained some experience in writing symmetries in standard form. Here is an exercise which concerns the symmetries of the parallelogram lattice.

This notation was developed in Section 5 of Unit IB4.

We use C_2 here rather than the notation R_r of Unit IB3, because the cyclic groups C_4 and C_6 are involved in the description of the symmetry groups of the square and hexagonal lattices.

Exercise 4.2

Write the composite $f_1 f_2 f_3$ of the symmetries $f_1 = r[\mathbf{a}, \pi]$, $f_2 = t[2\mathbf{b}]$ and $f_3 = r[\mathbf{b}, \pi]$ in the form xy, where $x \in T_2$ and $y \in C_2$.

For all the five types of lattice, we shall be looking at various group actions of $\Gamma(L)$. Since the elements of $\Gamma(L)$ are bijections of $\mathbb{R}^2$, there is certainly the group action where a group element g maps a point $\mathbf{x}$ to its image $g(\mathbf{x})$. If we take a subset X of $\mathbb{R}^2$ with the property that, for all g in $\Gamma(L)$, the point $g(\mathbf{x})$ belongs to X whenever $\mathbf{x}$ belongs to X, then there is a group action of $\Gamma(L)$ on X. We shall look at an example next.

Example 4.1

Let X be the set of all 2-centres of a parallelogram lattice $L = L\{\mathbf{a}, \mathbf{b}\}$. If g is any element of $\Gamma(L)$ and y is a rotation of L with centre $\mathbf{c}$, then the point $g(\mathbf{c})$ will be the centre of the rotation gyg^{-1}. Since $\mathbf{c}$ is a 2-centre, it follows that $g(\mathbf{c})$ is also a 2-centre. Hence $\Gamma(L)$ acts on the set X of all 2-centres.

Whenever we have a group action on a set X, it is useful to examine the orbits of the action and also to examine the stabilizers of various points of X. Let us consider the orbits first.

The 2-centres of L are at the points $\mathbf{c} = \tfrac{1}{2} n \mathbf{a} + \tfrac{1}{2} m \mathbf{b}$ $(n, m \in \mathbb{Z})$. An element g of $\Gamma(L)$ will be either a translation $t[\mathbf{d}]$ or a rotation $t[\mathbf{d}]\,r$, where $\mathbf{d} = p\mathbf{a} + q\mathbf{b}$ for some $p, q \in \mathbb{Z}$. Thus,

if $g = t[\mathbf{d}]$, then $g(\mathbf{c}) = \mathbf{c} + \mathbf{d}$
$$= \left(\tfrac{1}{2} n + p\right) \mathbf{a} + \left(\tfrac{1}{2} m + q\right) \mathbf{b}$$
$$= \tfrac{1}{2} n' \mathbf{a} + \tfrac{1}{2} m' \mathbf{b},$$

where $n' = n + 2p$ and $m' = m + 2q$;

if $g = t[\mathbf{d}]\,r$, then $g(\mathbf{c}) = -\mathbf{c} + \mathbf{d}$
$$= \left(-\tfrac{1}{2} n + p\right) \mathbf{a} + \left(-\tfrac{1}{2} m + q\right) \mathbf{b}$$
$$= \tfrac{1}{2} n' \mathbf{a} + \tfrac{1}{2} m' \mathbf{b},$$

where $n' = -n + 2p$ and $m' = -m + 2q$.

In either case, therefore, we have

$$g \left(\tfrac{1}{2} n \mathbf{a} + \tfrac{1}{2} m \mathbf{b}\right) = \tfrac{1}{2} n' \mathbf{a} + \tfrac{1}{2} m' \mathbf{b},$$

where n' has the same parity as n, and m' has the same parity as m. That is to say:

- if n and m are both even, so are n' and m';
- if n is odd and m is even, then n' is odd and m' even;
- if n is even and m is odd, then n' is even and m' odd;
- if n and m are both odd, so are n' and m'.

This shows that the 2-centres $\frac{1}{2}n\mathbf{a} + \frac{1}{2}m\mathbf{b}$ ($n, m \in \mathbb{Z}$) are partitioned into four orbits under the action of $\Gamma(L)$, each orbit being determined by the parities of n and m.

A geometric way of viewing these orbits is as follows. We said in Section 1 that 'a lattice can be thought of as the set of crossing points of two sets of equally spaced parallel lines such as you might find in some lattice work or a garden trellis'. Now, if we actually draw in these lines, then we obtain a tiling $\mathcal{T}$ of the plane, each tile of which is a parallelogram, as shown in Figure 4.1.

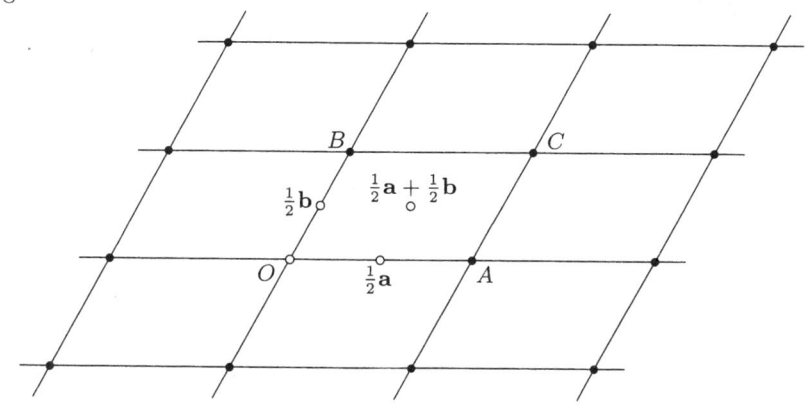

○ : representative of orbit

Figure 4.1

Now any element of $\Gamma(L)$ maps $\mathcal{T}$ to itself, and hence maps:

- the vertices of $\mathcal{T}$ to themselves;
- the centres of the parallelograms to themselves;
- the centres of the long edges to themselves;
- the centres of the short edges to themselves.

These sets of points are exactly the four orbits! ◇

Exercise 4.3

For each orbit defined in terms of the parities of n and m, determine the corresponding geometrically defined orbit.

Example 4.1 continued

In terms of the basic parallelogram $OABC$ in Figure 4.1, we may take the origin, the midpoints of OA and OB and the centre of $OABC$, as *representatives* of these four orbits. These representatives are shown as white dots in Figure 4.1; their position vectors are respectively $\mathbf{0}, \frac{1}{2}\mathbf{a}, \frac{1}{2}\mathbf{b}$ and $\frac{1}{2}\mathbf{a} + \frac{1}{2}\mathbf{b}$.

Next, we consider the stabilizers of the 2-centres.

Given any 2-centre whose position vector is $\mathbf{c}$, there are exactly two elements of $\Gamma(L)$ which fix $\mathbf{c}$, namely the identity element e and the rotation $r[\mathbf{c}, \pi] = t[2\mathbf{c}]\,r$. Thus,

$$\operatorname{Stab}(\mathbf{c}) = \{e, t[2\mathbf{c}]\,r\} \cong C_2.$$

This is true for each of the four orbits of 2-centres. You may be interested in comparing this analysis with Subsection 5.2 of *Unit GE2*, where we investigated tile and edge stabilizers of tilings. The parts of the tiling $\mathcal{T}$ in Figure 4.1 partition into four orbits (one tile orbit, one vertex orbit and two edge orbits), and each of these types of orbit also has a stabilizer equal to C_2. In fact, the stabilizer of a vertex V of $\mathcal{T}$ is exactly the same as the stabilizer of the 2-centre of L which occupies the same position as V; the stabilizer of the tile T of $\mathcal{T}$ is exactly the same as the stabilizer of the 2-centre of L which lies at the centre of T; and so on. ◆

4.2 The rectangular lattice

The symmetries of a rectangular lattice $L = L(\mathbf{a}, \mathbf{b})$ will include those of the parallelogram lattice. There will be the translations and rotations of order 2. Since there are no rotations with orders greater than 2, these comprise the direct symmetries of L. The subgroup $\Gamma^+(L)$ of direct symmetries is therefore

$$\Gamma^+(L) = \langle t_a, t_b, r : t_b t_a = t_a t_b,\ rt_a = t_a^{-1} r,\ rt_b = t_b^{-1} r,\ r^2 = e \rangle$$

where we write r for $r[\pi]$, as before.

In the shorter notation, we can write the group as

$$\Gamma^+(L) = \{xy : x \in T_2,\ y \in C_2;\ rx = x^{-1} r\}.$$

We have not yet needed to specify the basis $\{\mathbf{a}, \mathbf{b}\}$. The above remarks apply to any choice of basis for L. To proceed further, we take $\{\mathbf{a}, \mathbf{b}\}$ to be a reduced basis. This will ensure (in the case of the rectangular lattice) that there are reflections in the directions of t_a and t_b.

The lattice L has indirect symmetries. If the x-axis is chosen to contain the point $\mathbf{a}$, there will be the reflections $q[0]$ and $q[\pi/2]$ in axes through the origin O. All the indirect symmetries of L can be expressed as $t[\mathbf{d}]\,q[0]$ or as $t[\mathbf{d}]\,q[\pi/2]$, where $\mathbf{d}$ is a translation of L. For simplicity of notation, let us put $q = q[0]$ and $q' = q[\pi/2]$. The elements t_a, t_b, r, q and q' will certainly form a set of generators for the group $\Gamma(L)$, but we can manage with fewer. The composite qq' is the rotation r, so we could omit r and simply take the set $\{t_a, t_b, q, q'\}$. Alternatively, we can take the set $\{t_a, t_b, r, q\}$. this will be a set of generators for $\Gamma(L)$, since we can write $q' = qr$.

We already have the relations for dealing with elements which are composites of t_a, t_b and r and their inverses, so we may as well take the set $\{t_a, t_b, r, q\}$ as our set of generators.

Since the axis of q contains $\mathbf{a}$, we have $q(\mathbf{a}) = \mathbf{a}$. Using the Equation 6b of the Isometry Toolkit, we find that $q\,t[\mathbf{a}] = t[q(\mathbf{a})]\,q = t[\mathbf{a}]\,q$; that is, $qt_a = t_a q$.

Since $\mathbf{b}$ is perpendicular to the axis of q, we have $q(\mathbf{b}) = -\mathbf{b}$. It follows that $q\,t[\mathbf{b}] = t[q(\mathbf{b})]\,q = t[-\mathbf{b}]\,q = (t[\mathbf{b}])^{-1}\,q$; that is, $qt_b = t_b^{-1} q$.

We have found two of the relations that we need. Since q is a reflection, we must have $q^2 = e$. Finally, we notice that the reflection $q[\pi/2]$ can be expressed as either qr or as rq so we get $qr = rq$. In terms of these generators, every symmetry of the lattice can be written as one of the following expressions.

$\quad t_a^n t_b^m$ (translation)
$\quad t_a^n t_b^m r$ (rotation)
$\quad t_a^n t_b^m q$ (reflection or glide reflection)
$\quad t_a^n t_b^m rq$ (reflection or glide reflection)

You may write these, if you wish, as a single expression $t_a^n\, t_b^m\, r^\alpha\, q^\beta$ where α and β can take the values 0 and 1.

Our description of the group of symmetries of a rectangular lattice is thus

$$\Gamma(L) = \langle t_a, t_b, r, q : t_b t_a = t_a t_b,\ r t_a = t_a^{-1} r,\ r t_b = t_b^{-1} r,\ r^2 = e,$$
$$q t_a = t_a q,\ q t_b = t_b^{-1} q,\ qr = rq,\ q^2 = e \rangle.$$

Notice how these relations enable us to write every composite of symmetries in one of the forms listed above. They allow us to change the order in which the various generators occur in any expression.

The group $V = \{e, r, q, rq\}$ is the Klein group, and it has the relations $r^2 = e, q^2 = e$ and $qr = rq$. We write it here as D_2, since the groups D_4 and D_6 are involved in the descriptions of the symmetry groups of the square and hexagonal lattices. We may thus write the group $\Gamma(L)$ more simply as

$$\Gamma(L) = \{xy : x \in T_2,\ y \in D_2;\ r t_a = t_a^{-1} r,\ r t_b = t_b^{-1} r,$$
$$q t_a = t_a q,\ q t_b = t_b^{-1} q\}.$$

When writing $\Gamma(L)$ in this way, we put in only the relations which are needed to form composites of elements involving both T_2 and D_2. We assume the relations within each of these groups are known.

Here is an exercise in the use of these relations.

Exercise 4.4 _____

Write the composite $f_1 f_2 f_3$ of the symmetries

$$f_1 = r[\mathbf{a} + \mathbf{b}, \pi],\quad f_2 = q[\tfrac{1}{2}\mathbf{b}, 0] \quad \text{and} \quad f_3 = t[\mathbf{a}]$$

in the form xy, where $x \in T_2$ and $y \in D_2$.

_____

We now consider the action of $\Gamma(L)$ on the set of 2-centres. The following exercise concerns the orbits.

Exercise 4.5 _____

Find the orbits of the 2-centres of the rectangular lattice under the action of $\Gamma(L)$.

_____

In Exercise 4.5, you found that there are exactly four orbits of 2-centres. You will notice that these are the same orbits as for the parallelogram lattice. We now consider the stabilizers.

In Subsection 3.2, we showed that there are two reflection axes passing through each 2-centre. These, together with the identity and the rotation of order 2, will give us a group of order 4 as the stabilizer of any 2-centre. Since every element of this subgroup has order 2, the group is D_2, or the Klein group. Thus,

$$\text{Stab}(\mathbf{c}) = D_2 \quad (\mathbf{c} \text{ a 2-centre of } L).$$

Exercise 4.6 _____

For the 2-centre $\mathbf{c} = \tfrac{1}{2} n\mathbf{a} + \tfrac{1}{2} m\mathbf{b}$, write down explicitly the symmetries that fix $\mathbf{c}$.

_____

There is another group action which we shall examine. For any lattice L, the elements of $\Gamma(L)$ are affine transformations. Hence they map the set of lines in $\mathbb{R}^2$ to itself. It follows that $\Gamma(L)$ acts on the set of all lines in $\mathbb{R}^2$. We are going to consider the restriction of this action to lines which are reflection axes of L. This is a group action since, if l is the axis of a reflection q, then $g(l)$ will be the axis of the reflection $g q g^{-1}$ of L.

For the rectangular lattice L, let l be the axis of q and let l' be the axis of $q' = rq$. Now $\mathrm{Orb}(l) = \{g(l) : g \in G\}$ and $g(l)$ is the axis of the reflection gqg^{-1}. From our relations, we see that $gqg^{-1} = q$ when g is t_a, r or q. It follows that it will be sufficient to take $g = t_b^m$. The conjugate gqg^{-1} will then be $t_b^m q t_b^{-m} = t_b^{2m} q$. The axis of this reflection is $g(l)$, and it passes through the lattice point $m\mathbf{b}$. It follows that $\mathrm{Orb}(l)$ consists of all the axes parallel to l which pass through lattice points. In a similar way, we can find $\mathrm{Orb}(l')$.

Exercise 4.7

Show that $\mathrm{Orb}(l')$ consists of the lines parallel to l' which pass through the lattice points $n\mathbf{a}$.

There are two more orbits under this action. The axis of the reflection $t_b q$ is a line l_1 parallel to l which passes through the point $\tfrac{1}{2}\mathbf{b}$. The images $g(l_1)$ will be the lines parallel to l passing through the points $\left(n + \tfrac{1}{2}\right)\mathbf{b}$. The remaining orbit will consist of the lines parallel to l' which pass through the points $\left(m + \tfrac{1}{2}\right)\mathbf{a}$.

You will find that, once you have become familiar with the symmetries of a lattice, you will be able to write down the orbits of points and lines without needing to go through all the algebra. On the other hand, it is easy to make mistakes. The formal approach, using the defining relations to simplify conjugates, is the most reliable method.

Let us now look at the stabilizers of the reflection axes. We can see that the line l, which is the axis of q, will be fixed by e, r, q and rq. It is also fixed by the translations t_a^n and hence by the symmetries $t_a^n q$, $t_a^n r$ and $t_a^n rq$. Letting T_a be the group $\{t_a^n : n \in \mathbb{Z}\}$, the set of all these symmetries forms the group

$$\mathrm{Stab}(l) = \{xy : x \in T_a,\ y \in D_2;\ rt_a = t_a^{-1}r,\ qt_a = t_a q\}.$$

The relations here are those for the group $\Gamma(L)$ which do not involve the translations t_b. You may recognize this as the group $\Gamma(F_6)$ which is the symmetry group of a frieze of Type 6. Here we use q instead of h.

In *Unit IB3*, we also used the notations f_{vh} and $pmm2$ for this group.

Exercise 4.8

Find $\mathrm{Stab}(l')$, where l' is the axis of the reflection $q' = rq$.

4.3 The rhombic lattice

Let $L = L(\mathbf{a}, \mathbf{b})$ be a rhombic lattice. As with the rectangular lattice, $\Gamma(L)$ has a subgroup consisting of the symmetries of the parallelogram lattice; this subgroup is $\Gamma^+(L)$. So we have

$$\Gamma^+(L) = \{xy : x \in T_2,\ y \in C_2;\ rx = x^{-1}r\}.$$

We shall choose a basis $\{\mathbf{a}, \mathbf{b}\}$ for L for which $\|\mathbf{a}\| = \|\mathbf{b}\|$.

Then L has a reflection q in the line through $\mathbf{0}$ containing the point $\mathbf{a} + \mathbf{b}$, and a reflection $q' = rq$ in the line through $\mathbf{0}$ containing the point $\mathbf{b} - \mathbf{a}$. As in the case of the rectangular lattice, the symmetries e, r, q and rq form the group D_2, whose relations are

$$r^2 = e, \quad q^2 = e \quad \text{and} \quad qr = rq.$$

As we saw in Subsection 3.3, this is always possible for a rhombic lattice, but it is *not* always true in the rhombic case that a basis with this property is a reduced basis.

To specify $\Gamma(L)$, we need relations which connect the elements of T_2 and D_2. Since $q(\mathbf{a}) = \mathbf{b}$, it follows that $q\, t[\mathbf{a}] = t[q(\mathbf{a})]\, q = t[\mathbf{b}]\, q$; i.e.

$$qt_a = t_b q.$$

Now $q^2 = e$, so $\mathbf{a} = q(\mathbf{b})$, and thus we also have the relation

$$qt_b = t_a q.$$

Only one of these relations is needed since one follows from the other, but there is no harm in including both. Thus we can write $\Gamma(L)$ as

$$\Gamma(L) = \{xy : x \in T_2, y \in D_2; \; rx = x^{-1}r, \; qt_a = t_b q, \; qt_b = t_a q\}.$$

As in the case of the rectangular lattice, a general element g of $\Gamma(L)$ can be written as $g = t_a^n t_b^m r^\alpha q^\beta$, where α and β are 0 or 1.

Let us examine the action of $\Gamma(L)$ on the 2-centres of this lattice and find its orbits. We have seen that the 2-centres for every lattice $L(\mathbf{a}, \mathbf{b})$ are the points $\frac{1}{2}n\mathbf{a} + \frac{1}{2}m\mathbf{b}$. Now the group $\Gamma^+(L)$ is the same as the group of the parallelogram lattice, so it follows that there will be four orbits under the action of $\Gamma^+(L)$. The points $\mathbf{0}$, $\frac{1}{2}\mathbf{a}$, $\frac{1}{2}\mathbf{b}$ and $\frac{1}{2}\mathbf{a} + \frac{1}{2}\mathbf{b}$ will be representatives of these orbits. We need to see whether or not these four points belong to separate orbits under the action of $\Gamma(L)$. When we examine this, we find that the reflection q maps $\frac{1}{2}\mathbf{a}$ to $\frac{1}{2}\mathbf{b}$, so they will be in the same orbit. There are no elements of $\Gamma(L)$ which map $\mathbf{0}$ to $\frac{1}{2}\mathbf{a}$ or to $\frac{1}{2}\mathbf{a} + \frac{1}{2}\mathbf{b}$, and none which maps $\frac{1}{2}\mathbf{a}$ to $\frac{1}{2}\mathbf{a} + \frac{1}{2}\mathbf{b}$; so $\mathbf{0}$, $\frac{1}{2}\mathbf{a}$ and $\frac{1}{2}\mathbf{a} + \frac{1}{2}\mathbf{b}$ will be in separate orbits. Thus we have shown that there are exactly three orbits in this case.

Exercise 4.9

Find the stabilizers of the points $\mathbf{0}$, $\frac{1}{2}\mathbf{a}$ and $\frac{1}{2}\mathbf{a} + \frac{1}{2}\mathbf{b}$ under the action of $\Gamma(L)$.

Let us now look at the action of $\Gamma(L)$ on the axes of reflection. Through the point $\mathbf{0}$, there are lines l and l' which are the axes of q and $q' = rq$, respectively. The orbit of l consists of all the lines $g(l)$, where $g \in \Gamma(L)$. Through each lattice point there will be a reflection axis parallel to l, and these reflection axes will belong to one orbit. Similarly, there will be an orbit consisting of the axes which are parallel to l'. Thus, there are two orbits of reflection axes.

The stabilizer of l will contain the subgroup D_2 and will also contain the powers of the translation $t_a t_b$ parallel to l. Similarly, the stabilizer of l' will contain D_2 and all powers of the translation $t_a^{-1} t_b$ parallel to l'. Both these groups are isomorphic to the frieze group $\Gamma(F_6)$.

This lattice also has glide axes. These are shown as broken lines in Figure 3.6 of Subsection 3.3. Examples are the axes of the glide reflections $t_b q$ and $t_b r q$. It is not difficult to show that $\Gamma(L)$ acts on this set of lines, but we prefer to omit this and move on to another lattice.

4.4 The square lattice

As we saw earlier, a square lattice will have the symmetries of both the rectangular lattice and the rhombic lattice. In addition, it has rotations of order 4. Let $L = L(\mathbf{a}, \mathbf{b})$ be a square lattice with a reduced basis $\{\mathbf{a}, \mathbf{b}\}$, and let r be the rotation $r[\pi/2]$. The rotations which fix $\mathbf{0}$ form the cyclic group $C_4 = \langle r \rangle = \{e, r, r^2, r^3\}$. The rotations of L of order 4 are of the form $t_a^n t_b^m r$ or $t_a^n t_b^m r^3$, and the rotations of order 2 are of the form $t_a^n t_b^m r^2$. We may assume that the basis has been chosen so that $r(\mathbf{a}) = \mathbf{b}$ and $r(\mathbf{b}) = -\mathbf{a}$. This implies that we have the relations $rt_a = t_b r$ and $rt_b = t_a^{-1} r$. The group $\Gamma^+(L)$ of direct symmetries of the lattice can be written as

$$\Gamma^+(L) = \{xy : x \in T_2, y \in C_4; \; rt_a = t_b r, \; rt_b = t_a^{-1} r\}.$$

The lattice also has indirect symmetries. Let q be the reflection in the line through $\mathbf{0}$ containing the point $\mathbf{a}$. Then rq, $r^2 q$, $r^3 q$ will also be reflections with axes passing through $\mathbf{0}$.

The group of symmetries which fix $\mathbf{0}$ is generated by the set $\{r, q\}$. This group is a group of order 8, known as D_4, and it was discussed in *Unit IB4*, Subsection 3.2. Apart from the relations $r^4 = e$ and $q^2 = e$, we have the relation $qr = r^3 q = r^{-1} q$. In terms of generators and relations this group can be written as

$$D_4 = \langle r, q \colon r^4 = e,\ q^2 = e,\ qr = r^{-1} q \rangle.$$

Before we can fully specify the group $\Gamma(L)$, we need to relate q to the translations. Since $q(\mathbf{a}) = \mathbf{a}$ and $q(\mathbf{b}) = -\mathbf{b}$, we can obtain the relations $qt_a = t_a q$ and $qt_b = t_b^{-1} q$ as in the case of the rectangular lattice. We may now write $\Gamma(L)$ as

$$\Gamma(L) = \{xy \colon x \in T_2,\ y \in D_4;\ rt_a = t_b r,\ rt_b = t_a^{-1} r,$$
$$qt_a = t_a q,\ qt_b = t_b^{-1} q\}.$$

In Subsection 3.4, we found all the 4-centres of the square lattice. In the corresponding square tiling, they are the vertices and centres of each tile. We can show that $\Gamma(L)$ acts on the set of 4-centres; we leave this as an exercise.

Exercise 4.10

Show that $\Gamma(L)$ acts on the set of 4-centres of a square lattice L. Find the orbits under this action and the stabilizers of a point from each orbit.

Exercise 4.11

Show that $\Gamma(L)$ acts on the set of 2-centres which are not 4-centres. Show that they form a single orbit, and find the stabilizer of a point in the orbit.

4.5 The hexagonal lattice

The symmetry group $\Gamma(L)$ of a hexagonal lattice $L = L(\mathbf{a}, \mathbf{b})$ has rotations of order 6, so the rotation $r[\pi/3]$ will belong to $\Gamma(L)$. Let us put $r = r[\pi/3]$ and choose a reduced basis $\{\mathbf{a}, \mathbf{b}\}$ so that $r(\mathbf{a}) = \mathbf{b}$ and $r(\mathbf{b}) = \mathbf{b} - \mathbf{a}$. From these, we get the relations $rt_a = t_b r$ and $rt_b = t_a^{-1} t_b r$. The cyclic group C_6 generated by r has order 6, and we can write the group $\Gamma^+(L)$ as

$$\Gamma^+(L) = \{xy \colon x \in T_2,\ y \in C_6;\ rt_a = t_b r,\ rt_b = t_a^{-1} t_b r\}.$$

Let q be the reflection in the line through the origin and the point $\mathbf{a}$. We find that $q(\mathbf{a}) = \mathbf{a}$ and $q(\mathbf{b}) = \mathbf{a} - \mathbf{b}$. This implies that $qt_a = t_a q$ and $qt_b = t_a t_b^{-1} q$. We also have relation $qr = r^{-1} q$. The group of symmetries of L which fix $\mathbf{0}$ is the group D_6, which we can write as

$$D_6 = \langle r, q \colon r^6 = e,\ q^2 = e,\ qr = r^{-1} q \rangle.$$

The group $\Gamma(L)$ is then

$$\Gamma(L) = \{xy \colon x \in T_2,\ y \in D_6;\ rt_a = t_b r,\ rt_b = t_a^{-1} t_b r,$$
$$qt_a = t_a q,\ qt_b = t_a t_b^{-1} q\}.$$

In this lattice, there are 2-centres, 3-centres and 6-centres. The group $\Gamma(L)$ will act on the set of all centres of rotation. If $\mathbf{c}$ is the centre of a rotation r' of order n, then $f(\mathbf{c})$, where f is some symmetry of L, will be the centre of the rotation $f r' f^{-1}$, which also has order n.

Exercise 4.12

Find the orbits and the stabilizer of a point in each orbit for the action of $\Gamma(L)$ on centres of rotation.

5 THE CLASSIFICATION OF PLANE LATTICES

Before we return to the business of showing that there are just five types of plane lattice and no more, it is worth revisiting the minimality conditions, discussed in Section 1, and giving a more geometric description of them.

5.1 The minimality conditions revisited

In Theorem 1.6, we showed that a pair $\{\mathbf{a}, \mathbf{b}\}$ of vectors in a lattice L, satisfying the minimality conditions, will be a basis for L and will also satisfy the following three conditions, in which θ is the angle between $\mathbf{a}$ and $\mathbf{b}$:

(a) $\|\mathbf{a}\| \leq \|\mathbf{b}\|$;
(b) $-\|\mathbf{a}\|/2 \leq \|\mathbf{b}\| \cos \theta \leq \|\mathbf{a}\|/2$;
(c) $\pi/3 \leq \theta \leq 2\pi/3$.

Exercise 5.1

Show that, if $\{\mathbf{a}, \mathbf{b}\}$ obeys Conditions (a) and (b), then the angle θ obeys Condition (c).

We may interpret Conditions (a) and (b) geometrically, by choosing a coordinate system in which $\mathbf{a}$ is of unit length in the positive direction of the x-axis; that is,

$$\mathbf{a} = (1, 0).$$

Then Condition (a) states that $\mathbf{b}$ lies on or outside the unit circle $x^2 + y^2 = 1$. Furthermore, since $\mathbf{b} = (\|\mathbf{b}\| \cos \theta, \|\mathbf{b}\| \sin \theta)$, Condition (b) states that $\mathbf{b}$ lies between the two lines $x = -\frac{1}{2}$, $x = \frac{1}{2}$. Thus, Conditions (a) and (b) together state that $\mathbf{b}$ must lie in the shaded region shown in Figure 5.1.

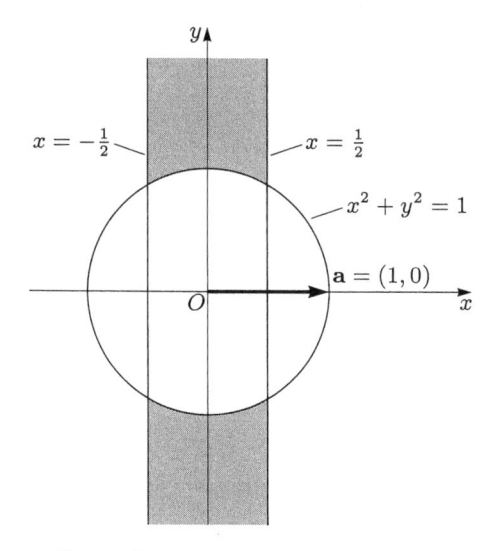

Figure 5.1

This is an important enough region to deserve a formal definition.

> **Definition 5.1 Minimality region**
>
> The **minimality region** is the region of $\mathbb{R}^2$ obeying the inequalities:
> (a) $x^2 + y^2 \geq 1$;
> (b) $-\frac{1}{2} \leq x \leq \frac{1}{2}$.

We can now state and prove the following theorem.

> **Theorem 5.1**
>
> Let $\{\mathbf{a}, \mathbf{b}\}$ be a reduced basis for a plane lattice L, and choose a coordinate system in which $\mathbf{a} = (1, 0)$. Then $\mathbf{b}$ lies in the minimality region.

Proof

By the definition of a reduced basis, $\{\mathbf{a}, \mathbf{b}\}$ satisfies the minimality conditions. Therefore the pair obeys Conditions (a) and (b) of Theorem 1.6, and so $\mathbf{b}$ lies in the minimality region. ∎

Exercise 5.2

Show that it is always possible to find a reduced basis such that $\mathbf{b}$ lies in the 'upper' half of the minimality region — that is, the half with $y \geq 0$.

One more piece of preparatory work is needed before we tackle the classification theorem for plane lattices. This consists in looking in more detail at the indirect symmetries that occur in rectangular and rhombic lattices.

5.2 Rectangular and rhombic indirect symmetries

You studied the symmetries of the rectangular and rhombic lattices in Subsections 4.2 and 4.3, respectively. The descriptions of the symmetry groups immediately imply the following facts.

- If L is a rectangular lattice, then it has a basis $\{\mathbf{a}, \mathbf{b}\}$ and a reflection q (with axis through the origin) such that

    $$q(\mathbf{a}) = \mathbf{a}, \quad q(\mathbf{b}) = -\mathbf{b}.$$

- If L is a rhombic lattice, then it has a basis $\{\mathbf{a}, \mathbf{b}\}$ and a reflection q (with axis through the origin) such that

    $$q(\mathbf{a}) = \mathbf{b}, \quad q(\mathbf{b}) = \mathbf{a}.$$

A surprising fact about plane lattices is that, if f is *any* indirect symmetry of *any* lattice, then its linear part obeys exactly one of the above properties. In order to state this more precisely, we now make the following definition.

> *Definition 5.2 Rectangular and rhombic indirect symmetries.*
>
> Let L be any plane lattice, let f be any indirect symmetry of L, and let q be the linear part of f. Then:
>
> - if, for some basis $\{\mathbf{a}, \mathbf{b}\}$ of L,
>
> $$q(\mathbf{a}) = \mathbf{a}, \quad q(\mathbf{b}) = -\mathbf{b},$$
>
>   the symmetry f is said to be **rectangular**;
>
> - if, for some basis $\{\mathbf{a}, \mathbf{b}\}$ of L,
>
> $$q(\mathbf{a}) = \mathbf{b}, \quad q(\mathbf{b}) = \mathbf{a},$$
>
>   the symmetry f is said to be **rhombic**.

Our 'surprising fact' can now be stated as follows.

> **Theorem 5.2**
>
> Every indirect symmetry of a plane lattice is either rectangular or rhombic but never both.

Proof

Let f be an indirect symmetry of a plane lattice L. Let q be the linear part of f. Then q is a reflection in a line through the origin, and by Theorem 2.2, q is a symmetry of L.

Let $\mathbf{a}$ be a vector of least magnitude in L. Then $q(\mathbf{a})$ is also a point of the lattice L, and $\|q(\mathbf{a})\| = \|\mathbf{a}\|$. There are three possibilities: $q(\mathbf{a})$ is not a multiple of $\mathbf{a}$, $q(\mathbf{a}) = \mathbf{a}$, or $q(\mathbf{a}) = -\mathbf{a}$. We consider these separately.

Case 1 $q(\mathbf{a})$ *is not a multiple of* $\mathbf{a}$.

In this case, $\mathbf{a}$ and $q(\mathbf{a})$ are linearly independent. Thus (as they are each of least magnitude in L) $\{\mathbf{a}, q(\mathbf{a})\}$ is a reduced basis for L.

Set $q(\mathbf{a}) = \mathbf{b}$. Then

$$q(\mathbf{b}) = q^2(\mathbf{a})$$
$$= \mathbf{a},$$

and so f is rhombic (see Figure 5.2).

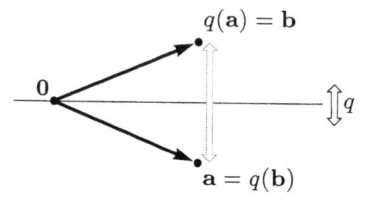

Figure 5.2

Case 2 $q(\mathbf{a}) = \mathbf{a}$.

Choose a coordinate system in which $\mathbf{a} = (1, 0)$, then choose $\mathbf{b} = (x, y)$ such that $\{\mathbf{a}, \mathbf{b}\}$ forms a reduced basis for L (see Figure 5.3). Then, by Theorem 5.1, $\mathbf{b}$ lies in the minimality region.

Let $\mathbf{c} = \mathbf{b} + q(\mathbf{b})$; then $\mathbf{c}$ belongs to L.

Since q is reflection in the x-axis, we have

$$q(\mathbf{b}) = (x, -y);$$
$$\mathbf{c} = (2x, 0).$$

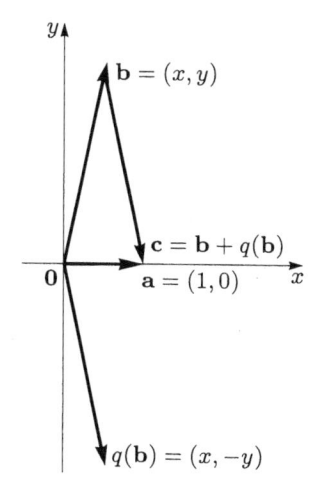

Figure 5.3

□

Exercise 5.3

Use the way in which $\mathbf{a}$ was chosen, to show that $\mathbf{c}$ must be $\mathbf{0}$ or $\pm\mathbf{a}$.

Proof of Theorem 5.2 continued

As a result of Exercise 5.3, we may now consider three subcases.

Subcase (a) $\mathbf{c} = \mathbf{0}$.

In this case, since $\mathbf{c}$ was defined as $\mathbf{b} + q(\mathbf{b})$, we have

$$q(\mathbf{b}) = -\mathbf{b}.$$

Since we are dealing with the case in which $q(\mathbf{a}) = \mathbf{a}$, we conclude that f is rectangular (see Figure 5.4).

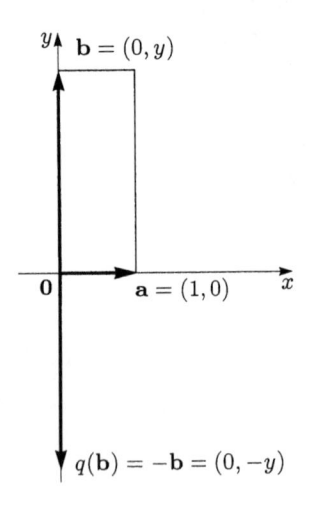

Figure 5.4

Subcase (b) $\mathbf{c} = \mathbf{a}$.

In this case,

$$\mathbf{b} + q(\mathbf{b}) = \mathbf{a},$$

and so

$$q(\mathbf{b}) = \mathbf{a} - \mathbf{b}.$$

Now the transition matrix from $\{\mathbf{b}, \mathbf{a} - \mathbf{b}\}$ to $\{\mathbf{a}, \mathbf{b}\}$ has integer entries and determinant -1, so by Theorem 1.2, $\{\mathbf{b}, \mathbf{a} - \mathbf{b}\}$ is a basis for L. Moreover,

$$q(\mathbf{b}) = \mathbf{a} - \mathbf{b}, \quad q(\mathbf{a} - \mathbf{b}) = q(\mathbf{a}) - (\mathbf{a} - \mathbf{b})$$
$$= \mathbf{a} - (\mathbf{a} - \mathbf{b})$$
$$= \mathbf{b},$$

and so, using this basis, we see that f is rhombic (see Figure 5.5).

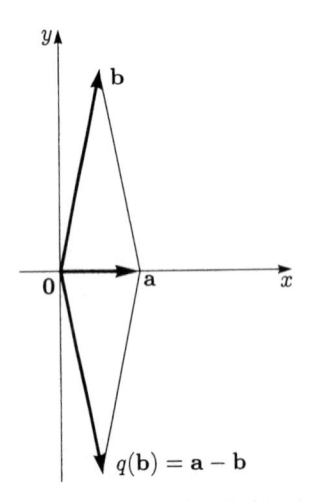

Figure 5.5

Subcase (c) **c** = −**a**.

In this case, a similar argument, using $\{-\mathbf{b}, \mathbf{a}+\mathbf{b}\}$, again shows that f is rhombic (see Figure 5.6).

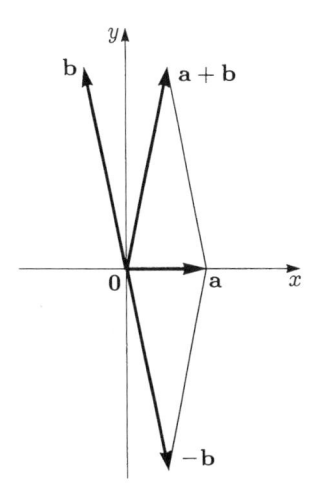

Figure 5.6

Exercise 5.4

Verify Subcase (c).

Proof of Theorem 5.2 continued

Case 3 $q(\mathbf{a}) = -\mathbf{a}$.

Again, choose a coordinate system in which $\mathbf{a} = (1,0)$, then choose $\mathbf{b} = (x,y)$ such that $\{\mathbf{a}, \mathbf{b}\}$ forms a reduced basis for L. Let $\mathbf{d} = \mathbf{b} - q(\mathbf{b})$. Since q is now reflection in the y-axis, we have

$$q(\mathbf{b}) = (-x, y);$$
$$\mathbf{d} = (2x, 0).$$

Exercise 5.5

Go through an argument similar to that of Case 2, and show that f is either rectangular or rhombic.

Proof of Theorem 5.2 continued

All that remains is to show that f cannot be both rectangular and rhombic. We do this by assuming that f is rectangular, and showing that it cannot also be rhombic.

Let q be the linear part of f. Then there is a basis $\{\mathbf{a}, \mathbf{b}\}$ for L such that

$$q(\mathbf{a}) = \mathbf{a}, \quad q(\mathbf{b}) = -\mathbf{b}.$$

Consider any vector $\mathbf{c} = n\mathbf{a} + m\mathbf{b}$ of L; thus, n and m are integers. We shall show that $\{\mathbf{c}, q(\mathbf{c})\}$ cannot be a basis for L.

Since q is linear, we have

$$q(\mathbf{c}) = n\mathbf{a} - m\mathbf{b},$$

and so the transition matrix from $\{\mathbf{c}, q(\mathbf{c})\}$ to $\{\mathbf{a}, \mathbf{b}\}$ is

$$\begin{bmatrix} n & n \\ m & -m \end{bmatrix},$$

whose determinant is $-2mn$. This cannot equal ± 1, and so (by Theorem 1.2) $\{\mathbf{c}, q(\mathbf{c})\}$ is not a basis for L. Thus there is no basis for L that makes f rhombic. ∎

5.3 The classification theorem

We are now (at last!) ready to prove that the five plane lattice types which we examined in Sections 3 and 4 are indeed the only geometric types of plane lattice that can occur. We do this by presenting a classification theorem that also summarizes their properties.

Theorem 5.3 Classification of plane lattices

The following five geometric types of plane lattice L are distinct from each other, and every plane lattice is of one of these types.

(a) The *parallelogram lattice*. In this case, for *every* basis $\{\mathbf{a},\mathbf{b}\}$ of L (including a reduced basis), $\|\mathbf{a}\| \neq \|\mathbf{b}\|$, and $\mathbf{a}$ and $\mathbf{b}$ are not orthogonal. The only symmetries of L are translations $t[n\mathbf{a}+m\mathbf{b}]$, and rotations of order 2 with centres $\frac{1}{2}(n\mathbf{a}+m\mathbf{b})$ $(n,m \in \mathbb{Z})$.

(b) The *rectangular lattice*. In this case, any reduced basis $\{\mathbf{a},\mathbf{b}\}$ of L has $\mathbf{a}$ and $\mathbf{b}$ orthogonal, and $\|\mathbf{a}\| \neq \|\mathbf{b}\|$. L has all the symmetries of a parallelogram lattice and, in addition, axes of reflection parallel to the directions of $\mathbf{a}$ and $\mathbf{b}$.

(c) The *rhombic lattice*. In this case, there is a basis $\{\mathbf{a},\mathbf{b}\}$ of L with $\|\mathbf{a}\| = \|\mathbf{b}\|$, but $\mathbf{a}$ and $\mathbf{b}$ are not orthogonal. L has all the symmetries of a parallelogram lattice and, in addition, axes of reflection and of essential glide reflection bisecting the angles between the directions of $\mathbf{a}$ and $\mathbf{b}$.

(d) The *square lattice*. In this case, any reduced basis $\{\mathbf{a},\mathbf{b}\}$ of L has $\|\mathbf{a}\| = \|\mathbf{b}\|$, and $\mathbf{a}$ and $\mathbf{b}$ are orthogonal. L has all the symmetries of a parallelogram lattice. In addition, it has axes of reflection parallel to the directions of $\mathbf{a}$ and $\mathbf{b}$, and axes of reflection and of essential glide reflection bisecting the angles between the directions of $\mathbf{a}$ and $\mathbf{b}$. Finally, it has rotation centres of order 4 at the points $\frac{1}{2}(n\mathbf{a}+m\mathbf{b})$ $(n,m \in \mathbb{Z}, n+m \in 2\mathbb{Z})$.

(e) The *hexagonal lattice*. In this case, L has a reduced basis $\{\mathbf{a},\mathbf{b}\}$ where $\|\mathbf{a}\| = \|\mathbf{b}\|$ and the angle between $\mathbf{a}$ and $\mathbf{b}$ is $\pi/3$. L has all the symmetries of a parallelogram lattice. In addition, it has axes of reflection and of essential glide reflection both parallel to and orthogonal to the directions of $\mathbf{a}, \mathbf{b}$ and $\mathbf{a} - \mathbf{b}$. Finally, it has rotation centres of order 6 at the points $n\mathbf{a}+m\mathbf{b}$ $(n,m \in \mathbb{Z})$ and of order 3 at the points $\frac{1}{3}(n\mathbf{a}+m\mathbf{b})$ $(n,m \in \mathbb{Z})$.

Proof

By Theorem 2.3, there are just three possibilities for the orders of the rotational symmetries of L: namely, $\{1,2\}, \{1,2,4\}$ and $\{1,2,3,6\}$. We deal with these cases in turn.

Case 1 L has only rotations of orders 1, 2.

Either L has no indirect symmetries, or it does have indirect symmetries. We now consider these two subcases.

Subcase (a) L has no indirect symmetries.

Let $\{\mathbf{a},\mathbf{b}\}$ be a reduced basis for L, and let $r = r[\pi]$. Then the group of symmetries of L that fix O is exactly $\{e,r\}$, which is generated by r. Since every symmetry of L is the composite of a symmetry that fixes O with a translational symmetry, it follows that

$$\Gamma(L) = \langle t_a, t_b, r \rangle.$$

This is exactly the group described in Subsection 4.1, and so the parallelogram lattice is the only possibility for this subcase.

Subcase (b) L has indirect symmetries.

Let $r = r[\pi]$, let f be any indirect symmetry of L, and let q be the linear part of f.

By Theorem 5.1, f (and hence q) is either rectangular or rhombic but not both.

Now, since r is the only non-trivial rotation about O, q and rq are the only reflections in axes through O. We now show that rq is of the same type (rectangular or rhombic) as q.

Suppose q is rectangular; then there is a basis $\{\mathbf{a}, \mathbf{b}\}$ such that

$$q(\mathbf{a}) = \mathbf{a}, \quad q(\mathbf{b}) = -\mathbf{b}.$$

Now

$$rq(\mathbf{a}) = r(\mathbf{a}) = -\mathbf{a},$$
$$rq(\mathbf{b}) = r(-\mathbf{b}) = \mathbf{b}.$$

Thus the basis $\{\mathbf{b}, \mathbf{a}\}$ shows that rq is rectangular.

Suppose q is rhombic; then there is a basis $\{\mathbf{a}, \mathbf{b}\}$ such that

$$q(\mathbf{a}) = \mathbf{b}, \quad q(\mathbf{b}) = \mathbf{a}$$

Now

$$rq(\mathbf{a}) = r(\mathbf{b}) = -\mathbf{b},$$
$$rq(\mathbf{b}) = r(\mathbf{a}) = -\mathbf{a}.$$

Thus, considering the basis $\{\mathbf{a}, -\mathbf{b}\}$, we have

$$rq(\mathbf{a}) = -\mathbf{b}, \quad rq(-\mathbf{b}) = \mathbf{a}.$$

Therefore rq is rhombic.

Thus the indirect symmetries of L must be either all rectangular (showing that L must be a rectangular lattice) or all rhombic (showing that L must be a rhombic lattice).

Case 2 L has rotations of orders 1, 2, 4

We proved in Subsection 3.4 that, if L has a rotation of order 4, then it has a reduced basis $\{\mathbf{a}, \mathbf{b}\}$ with $\|\mathbf{a}\| = \|\mathbf{b}\|$ and with $\mathbf{a}$ and $\mathbf{b}$ orthogonal.

Let $r = r[\pi/2]$ and let q be the reflection in the line through O containing the point $\mathbf{a}$. We saw in Subsection 4.4 that r and q generate the group of symmetries of L that fix O. Since every symmetry of L is the composite of a symmetry that fixes O with a translational symmetry, it follows that

$$\Gamma(L) = \langle t_a, t_b, r, q \rangle.$$

This is exactly the group described in Subsection 4.4, and so the square lattice is the only possibility for a lattice whose rotations are of orders 1, 2, 4.

Case 3 L has rotations of orders 1, 2, 3, 6

We proved in Subsection 3.5 that, if L has a rotation of order 3 (and hence also a rotation of order 6), then it has a reduced basis $\{\mathbf{a}, \mathbf{b}\}$ with $\|\mathbf{a}\| = \|\mathbf{b}\|$ where the angle between $\mathbf{a}$ and $\mathbf{b}$ is $\pi/3$.

Let $r = r[\pi/3]$ and let q be the reflection in the line through O containing the point $\mathbf{a}$. We saw in Subsection 4.5 that r and q generate the group of symmetries of L that fix O. Arguing just as in Case 2, we see that

$$\Gamma(L) = \langle t_a, t_b, r, q \rangle$$

is exactly the group described in Subsection 4.5, so the hexagonal lattice is the only possibility for a lattice whose rotations are of orders 1, 2, 3, 6. ∎

As a parting shot, we shall now show (without proof) how to use the minimality region to classify plane lattices in a more geometric spirit.

Let L be a plane lattice, with a reduced basis $\{\mathbf{a}, \mathbf{b}\}$. Choose a coordinate system in which $\mathbf{a} = (1, 0)$. By the result of Exercise 5.2, we may assume that $\mathbf{b}$ lies in the upper half of the minimality region; call this region R (see Figure 5.7).

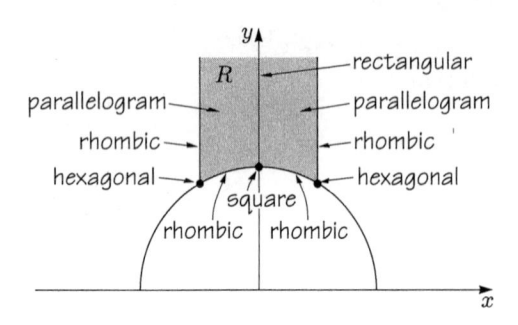

Figure 5.7

The following cases arise.

(a) If $\mathbf{b}$ lies in the interior of R, then L is a parallelogram lattice.

(b) If $\mathbf{b}$ lies on the y-axis, but not at $(0,1)$, then L is a rectangular lattice.

(c) If $\mathbf{b}$ lies on the circle $x^2 + y^2 = 1$, but not at $\left(\pm\frac{1}{2}, \frac{\sqrt{3}}{2}\right)$ or at $(0,1)$, then L is a rhombic lattice. Moreover, the reduced basis $\{\mathbf{a}, \mathbf{b}\}$ has the property that $q(\mathbf{a}) = \mathbf{b}, q(\mathbf{b}) = \mathbf{a}$ for some reflection of L.

(d) If $\mathbf{b}$ lies on one of the lines $x = \pm\frac{1}{2}$, but not at $\left(\pm\frac{1}{2}, \frac{\sqrt{3}}{2}\right)$, then L is a rhombic lattice. However, the reduced basis $\{\mathbf{a}, \mathbf{b}\}$ is *not* the basis with the property $q(\mathbf{a}) = \mathbf{b}, q(\mathbf{b}) = \mathbf{a}$.

(e) If $\mathbf{b} = (0, 1)$ then L is the square lattice.

(f) If $\mathbf{b} = \left(\pm\frac{1}{2}, \frac{\sqrt{3}}{2}\right)$, then L is the hexagonal lattice; in this case, $\left(\frac{1}{2}, \frac{\sqrt{3}}{2}\right)$ and $\left(-\frac{1}{2}, \frac{\sqrt{3}}{2}\right)$ are both points of L.

SOLUTIONS TO THE EXERCISES

Solution 1.1

The vectors $\mathbf{p}$ and $\mathbf{q}$ can be expressed as $\mathbf{p} = n_1\mathbf{a} + m_1\mathbf{b}$ and $\mathbf{q} = n_2\mathbf{a} + m_2\mathbf{b}$, where n_1, n_2, m_1 and m_2 are integers. An integer combination of the vectors $\mathbf{p}$ and $\mathbf{q}$ can be written as

$$n\mathbf{p} + m\mathbf{q} = n(n_1\mathbf{a} + m_1\mathbf{b}) + m(n_2\mathbf{a} + m_2\mathbf{b})$$
$$= (nn_1 + mn_2)\mathbf{a} + (nm_1 + mm_2)\mathbf{b}.$$

This is also an integer combination of $\mathbf{a}$ and $\mathbf{b}$, and hence belongs to $L(\mathbf{a}, \mathbf{b})$.

Solution 1.2

For each of these choices, we shall try to express a general point (n, m) as $\alpha \mathbf{a} + \beta \mathbf{b}$, where α and β are integers.

(a) With $(n, m) = \alpha(1, 0) + \beta(2, 1)$, we find $\alpha = n - 2m$ and $\beta = m$. The values of α and β are always integers, so this choice of $\mathbf{a}$ and $\mathbf{b}$ is suitable.

(b) With $(n, m) = \alpha(1, 0) + \beta(1, 2)$, we find $\alpha = n - m/2$ and $\beta = m/2$. These are not always integers, so this choice of $\mathbf{a}$ and $\mathbf{b}$ will *not* be suitable.

(c) With $(n, m) = \alpha(0, 1) + \beta(-1, 3)$, we find $\alpha = 3n + m$ and $\beta = -n$. These are always integers, so this choice of $\mathbf{a}$ and $\mathbf{b}$ is suitable.

(d) With $(n, m) = \alpha(2, 3) + \beta(1, 1)$, we find $\alpha = -n + m$ and $\beta = 3n - 2m$. These are always integers, so this choice of $\mathbf{a}$ and $\mathbf{b}$ is suitable.

Solution 1.3

(a) $\mathbf{a}' = (1, 0) = \mathbf{a}$, and $\mathbf{b}' = (2, 1) = 2\mathbf{a} + \mathbf{b}$. The transition matrix from $\{\mathbf{a}', \mathbf{b}'\}$ to $\{\mathbf{a}, \mathbf{b}\}$ is $\begin{bmatrix} 1 & 2 \\ 0 & 1 \end{bmatrix}$ and we find that $D = 1$. Hence $\{\mathbf{a}', \mathbf{b}'\}$ is a basis for the lattice.

(b) The transition matrix from $\{\mathbf{a}', \mathbf{b}'\}$ to $\{\mathbf{a}, \mathbf{b}\}$ is $\begin{bmatrix} 1 & 1 \\ 0 & 2 \end{bmatrix}$ and we find that $D = 2$. Hence $\{\mathbf{a}', \mathbf{b}'\}$ is *not* a basis for the lattice.

(c) The transition matrix from $\{\mathbf{a}', \mathbf{b}'\}$ to $\{\mathbf{a}, \mathbf{b}\}$ is $\begin{bmatrix} 0 & -1 \\ 1 & 3 \end{bmatrix}$ and we find that $D = 1$. Hence $\{\mathbf{a}', \mathbf{b}'\}$ is a basis for the lattice.

(d) The transition matrix from $\{\mathbf{a}', \mathbf{b}'\}$ to $\{\mathbf{a}, \mathbf{b}\}$ is $\begin{bmatrix} 2 & 1 \\ 3 & 1 \end{bmatrix}$ and we find that $D = -1$. Hence $\{\mathbf{a}', \mathbf{b}'\}$ is a basis for the lattice.

Solution 1.4

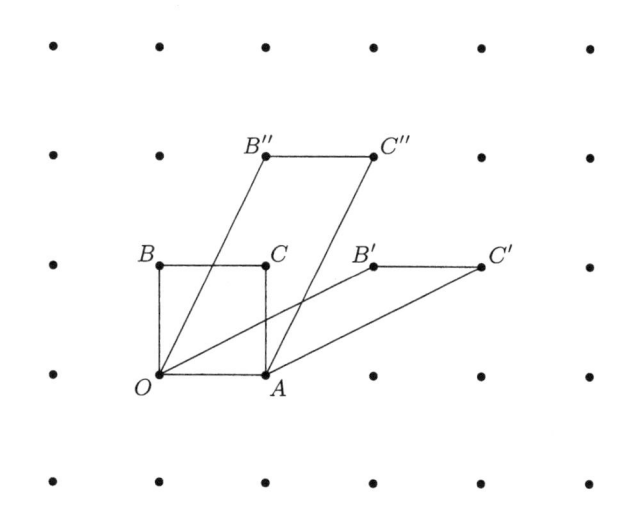

The areas of the parallelograms $OACB$, $OAC'B'$ and $OAC''B''$ are 1, 1 and 2, respectively. Only the parallelogram $OAC''B''$ contains a lattice point which is not one of the vertices.

Solution 1.5

The area is given by $|a_1 b_2 - a_2 b_1|$ in each case.

(a) 6

(b) 4

(c) $2\sqrt{3}$

Solution 1.6

Take any parallelogram whose vertices are points of a lattice $L = L(\mathbf{a}, \mathbf{b})$. We may, if necessary, apply a translation so that one of the vertices is the origin O. This does not change its area. Writing the parallelogram as $OA''C''B''$, we let $\mathbf{a}'' = (a_1'', a_2'')$ and $\mathbf{b}'' = (b_1'', b_2'')$ be the vectors corresponding to A'' and B''. We can then write $\mathbf{a}'' = t\mathbf{a} + u\mathbf{b}$ and $\mathbf{b}'' = v\mathbf{a} + w\mathbf{b}$, where $t, u, v,$ and w are integers. Putting these equations in matrix form, we get $\begin{bmatrix} a_1'' & a_2'' \\ b_1'' & b_2'' \end{bmatrix} = \begin{bmatrix} t & u \\ v & w \end{bmatrix} \begin{bmatrix} a_1 & a_2 \\ b_1 & b_2 \end{bmatrix}$.

Taking determinants of both sides and forming absolute values, we get $|a_1'' b_2'' - a_2'' b_1''| = |tw - uv||a_1 b_2 - a_2 b_1|$. The number $|tw - uv|$ is an integer, so we have shown that the area of $OA''C''B''$ is an integer multiple of the area of the basic parallelogram $OACB$.

Solution 1.7

We have
$$|\mathbf{a}| = \sqrt{3^2 + 4^2} = 5.$$

Thus there cannot be more than $\dfrac{10}{5} + 1 = 3$ lattice points of $\mathcal{L}_m$ in the disc, for any m.

We also have
$$|a_1 b_2 - a_2 b_1| = 7.$$

Thus there cannot be more than $\dfrac{10 \times 5}{7} + 1$ lines $\mathcal{L}_m$ intersecting the disc — i.e. no more than 8 lines, since the number of lines is an integer.

Therefore, there cannot be more than $3 \times 8 = 24$ lattice points in any disc of diameter 10.

Solution 1.8

The possible choices for $\mathbf{a}$ are $(0, 2)$ and $(0, -2)$. The possible choices for $\mathbf{b}$ are $(3, 0)$ and $(-3, 0)$.

Solution 1.9

The possible choices for $\mathbf{a}$ are again $(0, 2)$ and $(0, -2)$. The possible choices for $\mathbf{b}$ are $(2, 1), (2, -1), (-2, 1)$ and $(-2, -1)$.

Solution 2.1

(a) $g = q[(2,0),(0,1),0] = t[(2,2)]\,q[0]$ by Equation 15 of the Isometry Toolkit. Therefore the translation component is $t[(2,0)]$, the translation part is $t[(2,2)]$, the reflection component is $q[(0,1),0]$ and the linear part is $q[0]$.

(b) $q = q[(0,2),(0,1),\pi/2]$
$= q[(0,2),(0,0),\pi/2]$ (as $(0,0)$ and $(0,1)$ lie on the same vertical line)
$= t[(0,2)]\,q[\pi/2]$ (by Equation 15 of the Isometry Toolkit).

Therefore the translation component and the translation part are each equal to $t[(0,2)]$, while the reflection component and the linear part are each equal to $q[\pi/2]$.

(c) $g = q[(3,3),(2,0),\pi/4]$
$\quad = q[(3,3),(1,-1),\pi/4]$ (as $(2,0)$ and $(1,-1)$ lie on the same line inclined at angle $\pi/4$ to the x-axis)
$\quad = t[(5,1)]\, q\,[\pi/4]$ (by Equation 15 of the Isometry Toolkit).

Therefore the translation component is $t[(3,3)]$, the translation part is $t[(5,1)]$, the reflection component is $q[(1,-1),\pi/4]$ and the linear part is $q[\pi/4]$.

Solution 2.2

(a) is inessential, while (b) is essential.

Solution 2.3

Here $\mathbf{a} = (1,0)$ and $\mathbf{b} = (0,1)$. We find $r[\pi/2](\mathbf{a}) = \mathbf{b}$, $r[\pi/2](\mathbf{b}) = -\mathbf{a}$, $q[\pi/4](\mathbf{a}) = \mathbf{b}$ and $q[\pi/4](\mathbf{b}) = \mathbf{a}$.

(a) $f_1 f_2 = r[\pi/2]\, t[\mathbf{a} - \mathbf{b}]\, r[\pi/2]$
$\quad = t[\mathbf{b} + \mathbf{a}]\, r[\pi/2]\, r[\pi/2] = t[\mathbf{a} + \mathbf{b}]\, r[\pi].$

(b) $f_1 f_2 = q[\pi/4]\, t[\mathbf{a} + \mathbf{b}]\, q[3\pi/4]$
$\quad = t[\mathbf{b} + \mathbf{a}]\, q[\pi/4]\, q[3\pi/4] = t[\mathbf{a} + \mathbf{b}]\, r[\pi]$ (since $r[-\pi] = r[\pi]$).

(c) $f_1 f_2 = t[\mathbf{a}]\, r[\pi/2]\, t[\mathbf{b}]\, q[0]$
$\quad = t[\mathbf{a}]\, t[-\mathbf{a}]\, r[\pi/2]\, q[0] = q[\pi/4].$

(d) $f_1 f_2 = t[\mathbf{a}]\, q[\pi/4]\, t[\mathbf{a} - \mathbf{b}]\, r[\pi/2]$
$\quad = t[\mathbf{a}]\, t[\mathbf{b} - \mathbf{a}]\, q[\pi/4]\, r[\pi/2] = t[\mathbf{b}]\, q[0].$

Solution 2.4

Here $\mathbf{a} = (2,0)$ and $\mathbf{b} = (1,\sqrt{3})$. We find $r[\pi/3](\mathbf{a}) = \mathbf{b}$, $r[\pi/3](\mathbf{b}) = \mathbf{b} - \mathbf{a}$, $q[\pi/3](\mathbf{a}) = \mathbf{b} - \mathbf{a}$, $q[\pi/3](\mathbf{b}) = \mathbf{b}$, $q[\pi/6](\mathbf{a}) = \mathbf{b}$ and $q[\pi/6](\mathbf{b}) = \mathbf{a}$.

(a) $f_1 f_2 = r[\pi/3]\, t[\mathbf{a}]\, r[\pi/3]$
$\quad = t[\mathbf{b}]\, r[\pi/3]\, r[\pi/3] = t[\mathbf{b}]\, r[2\pi/3].$

(b) $f_1 f_2 = q[\pi/3]\, t[\mathbf{a} + \mathbf{b}]\, q[2\pi/3]$
$\quad = t[\mathbf{b} - \mathbf{a} + \mathbf{b}]\, q[\pi/3]\, q[2\pi/3] = t[2\mathbf{b} - \mathbf{a}]\, r[-2\pi/3] = t[2\mathbf{b} - \mathbf{a}]\, r[\pi/3].$

(c) $f_1 f_2 = t[\mathbf{a}]\, r[\pi/3]\, t[\mathbf{b}]\, q[0]$
$\quad = t[\mathbf{a}]\, t[\mathbf{b} - \mathbf{a}]\, r[\pi/3]\, q[0] = t[\mathbf{b}]\, q[\pi/6].$

(d) $f_1 f_2 = t[\mathbf{a}]\, q[\pi/6]\, t[\mathbf{b}]$
$\quad = t[\mathbf{a}]\, t[\mathbf{a}]\, q[\pi/6] = t[2\mathbf{a}]\, q[\pi/6].$

Solution 2.5

By Equation 8 of the Isometry Toolkit, the standard form for $r[\mathbf{c}, \theta]$ is $t[\mathbf{d}]\, r[\theta]$, where $\mathbf{d} = \mathbf{c} - r[\theta](\mathbf{c})$.

(a) We observe that $r[\pi/2](\mathbf{a}) = \mathbf{b}$ and $r[\pi/2](\mathbf{b}) = -\mathbf{a}$,
so $r[\pi/2]((\mathbf{a} + \mathbf{b})/2) = (\mathbf{b} - \mathbf{a})/2$.
Hence $\mathbf{d} = (\mathbf{a} + \mathbf{b})/2 - (\mathbf{b} - \mathbf{a})/2 = \mathbf{a}$,
so, in standard form, the rotation is $t[\mathbf{a}]\, r[\pi/2]$.

(b) Now $r[3\pi/2](\mathbf{a}) = -\mathbf{b}$ and $r[3\pi/2](\mathbf{b}) = \mathbf{a}$,
so $r[3\pi/2](\mathbf{a} + \mathbf{b}) = -\mathbf{b} + \mathbf{a}$.
Hence $\mathbf{d} = (\mathbf{a} + \mathbf{b}) - (-\mathbf{b} + \mathbf{a}) = 2\mathbf{b}$, so the standard form is $t[2\mathbf{b}]\, r[3\pi/2]$.

Solution 2.6

(a) We can see that $r[\pi/3](\mathbf{b}) = \mathbf{b} - \mathbf{a}$. Hence $\mathbf{d} = \mathbf{b} - (\mathbf{b} - \mathbf{a}) = \mathbf{a}$, so the standard form is $t[\mathbf{a}]\, r[\pi/3]$.

(b) Now $r[2\pi/3](\mathbf{a}) = \mathbf{b} - \mathbf{a}$ and $r[2\pi/3](\mathbf{b}) = -\mathbf{a}$, so we get
$r[2\pi/3]((\mathbf{a} + \mathbf{b})/3) = r[2\pi/3](\mathbf{a})/3 + r[2\pi/3](\mathbf{b})/3 = (\mathbf{b} - \mathbf{a})/3 - \mathbf{a}/3 = (\mathbf{b} - 2\mathbf{a})/3$.
Hence $\mathbf{d} = (\mathbf{a} + \mathbf{b})/3 - (\mathbf{b} - 2\mathbf{a})/3 = \mathbf{a}$, so the standard form is $t[\mathbf{a}]\, r[2\pi/3]$.

Solution 2.7

The standard form for $q[\mathbf{c}, \theta]$ is $t[\mathbf{d}] \, q[\theta]$, where $\mathbf{d} = \mathbf{c} - q[\theta](\mathbf{c})$.

(a) We have $q[\pi/2](\mathbf{a}) = -\mathbf{a}$ and $q[\pi/2](\mathbf{b}) = \mathbf{b}$,
so $q[\pi/2]((\mathbf{a}+\mathbf{b})/2) = (-\mathbf{a}+\mathbf{b})/2$.
Hence $\mathbf{d} = (\mathbf{a}+\mathbf{b})/2 - (-\mathbf{a}+\mathbf{b})/2 = \mathbf{a}$, so the reflection is $t[\mathbf{a}] \, q[\pi/2]$.

(b) We find that $q[3\pi/4](\mathbf{a}) = -\mathbf{b}$ and $q[3\pi/4](\mathbf{b}) = -\mathbf{a}$,
so $q[3\pi/4](\mathbf{a}+\mathbf{b}) = -\mathbf{b} - \mathbf{a}$.
Hence $\mathbf{d} = (\mathbf{a}+\mathbf{b}) - (-\mathbf{b} - \mathbf{a}) = 2(\mathbf{a}+\mathbf{b})$, so the reflection is
$t[2(\mathbf{a}+\mathbf{b})] \, q[3\pi/4]$.

Solution 2.8

(a) Now $q[\pi/6](\mathbf{b}) = \mathbf{a}$ so $\mathbf{d} = \mathbf{b} - \mathbf{a}$. The standard form of the reflection is
$t[\mathbf{b}-\mathbf{a}] \, q[\pi/6]$.

(b) Since $q[\pi/2](\mathbf{a}) = -\mathbf{a}$ and $q[\pi/2](\mathbf{b}) = \mathbf{b} - \mathbf{a}$,
we have $q[\pi/2]((\mathbf{a}+\mathbf{b})/3) = (-\mathbf{a}+\mathbf{b}-\mathbf{a})/3 = (\mathbf{b}-2\mathbf{a})/3$.
Hence $\mathbf{d} = (\mathbf{a}+\mathbf{b})/3 - (\mathbf{b}-2\mathbf{a})/3 = \mathbf{a}$, so the reflection is $t[\mathbf{a}] \, q[\pi/2]$.

Solution 2.9

(a) The translation component $t[\mathbf{g}] = t[\mathbf{a}]$ is a translation of the lattice, so the glide reflection is inessential.

We have
$$q[\mathbf{g}, \mathbf{c}, \theta] = q[\mathbf{a}, \mathbf{b}, 0]$$
$$= t[\mathbf{a}+2\mathbf{b}] \, q[0] \quad \text{(by Equation 15 of the Isometry Toolkit)},$$
and this is in standard form.

(b) The translation component $t[\mathbf{g}] = t[\tfrac{1}{2}\mathbf{a}+\tfrac{1}{2}\mathbf{b}]$ is not a translation of the lattice, so this glide reflection is essential.

We have
$$q[\mathbf{g}, \mathbf{c}, \theta] = q[\tfrac{1}{2}\mathbf{a}+\tfrac{1}{2}\mathbf{b}, \tfrac{1}{2}\mathbf{b}, \pi/4]$$
$$= t[\mathbf{d}] \, q[\pi/4] \quad \text{where } \mathbf{d} = \tfrac{1}{2}\mathbf{a}+\tfrac{1}{2}\mathbf{b}+\tfrac{1}{2}\mathbf{b}-\tfrac{1}{2}\mathbf{a}$$
$$= t[\mathbf{b}] \, q[\pi/4],$$

Using Equation 14 of the Isometry Toolkit, together with the fact that $q[\pi/4](\tfrac{1}{2}\mathbf{b}) = \tfrac{1}{2}\mathbf{a}$.

and this is in standard form.

Solution 2.10

(a) The translation component $t[\mathbf{g}] = t[\mathbf{b}]$ is a translation of the lattice, so the glide reflection is inessential.

We have
$$q[\mathbf{g}, \mathbf{c}, \theta] = q[\mathbf{b}, \mathbf{a}, \pi/3]$$
$$= t[\mathbf{b}+\mathbf{a}-q[\pi/3](\mathbf{a})] \, q[\pi/3] \quad \text{(by Equation 14 of the Isometry Toolkit)}$$
$$= t[\mathbf{b}+\mathbf{a}-(\mathbf{b}-\mathbf{a})] \, q[\pi/3]$$
$$= t[2\mathbf{a}] \, q[\pi/3],$$
and this is in standard form.

(b) The translation component $t[\mathbf{g}] = t[\tfrac{1}{2}\mathbf{a}]$ is not a translation of the lattice, so the glide reflection is essential.

We have
$$q[\mathbf{g}, \mathbf{c}, \theta] = q[\tfrac{1}{2}\mathbf{a}, \tfrac{1}{2}\mathbf{b}, 0]$$
$$= t[\tfrac{1}{2}\mathbf{a}+\tfrac{1}{2}\mathbf{b}-q[0](\tfrac{1}{2}\mathbf{b})] \, q[0] \quad \text{(by Equation 14 of the Isometry Toolkit)}$$
$$= t[\tfrac{1}{2}\mathbf{a}+\tfrac{1}{2}\mathbf{b}-\tfrac{1}{2}(\mathbf{a}-\mathbf{b})] \, q[0]$$
$$= t[\mathbf{b}] \, q[0],$$
and this is in standard form.

Solution 3.1

In each case, Equation 9 of the Isometry Toolkit tells us that the centre of rotation is at $\mathbf{c} = \frac{1}{2}\mathbf{d}$. Thus:

(a) the centre of rotation is at $\frac{1}{2}\mathbf{a}$;

(b) the centre of rotation is at $\frac{1}{2}\mathbf{a} + \frac{1}{2}\mathbf{b}$;

(c) the centre of rotation is at $\frac{1}{2}n\mathbf{a} + \frac{1}{2}m\mathbf{b}$.

Solution 3.2

$q_1 = t[\mathbf{b}]\, q;\quad q_2 = (t[\mathbf{b}])^2\, q\ (= t[2\mathbf{b}]\, q);\quad q' = r[\pi]\, q;$

$q_1' = t[\mathbf{a}]\, q' = t[\mathbf{a}]\, r[\pi]\, q;\quad q_2' = (t[\mathbf{a}])^2\, q' = (t[\mathbf{a}])^2\, r[\pi]\, q\ (= t[2\mathbf{a}]\, r[\pi]\, q).$

Solution 3.3

$q_1 = t[\mathbf{b}]\, t[\mathbf{a}]^{-1}\, q\ (= t[\mathbf{b} - \mathbf{a}]\, q);\quad q' = r[\pi]\, q;$

$q_1' = t[\mathbf{a}]\, t[\mathbf{b}]\, q' = t[\mathbf{a}]\, t[\mathbf{b}]\, r[\pi]\, q\ (= t[\mathbf{a} + \mathbf{b}]\, r[\pi]\, q);$

$g = t[\mathbf{b}]\, q;\quad g' = t[\mathbf{b}]\, q' = t[\mathbf{b}]\, r[\pi]\, q.$

Solution 3.4

By substituting the coordinates of $\mathbf{d}$ into the matrix equation

$\begin{bmatrix} c_1 \\ c_2 \end{bmatrix} = \begin{bmatrix} \frac{1}{2} & -\frac{1}{2} \\ \frac{1}{2} & \frac{1}{2} \end{bmatrix} \begin{bmatrix} d_1 \\ d_2 \end{bmatrix}$ in each case, we find the centre $\mathbf{c}$:

(a) $\mathbf{c} = \frac{1}{2}\mathbf{a} + \frac{1}{2}\mathbf{b}$;

(b) $\mathbf{c} = -\frac{1}{2}\mathbf{a} + \frac{1}{2}\mathbf{b}$;

(c) $\mathbf{c} = \mathbf{b}$.

Solution 3.5

The matrix representing $f[3\pi/2]$ with respect to $\mathbf{a}$ and $\mathbf{b}$ is $\begin{bmatrix} \frac{1}{2} & \frac{1}{2} \\ -\frac{1}{2} & \frac{1}{2} \end{bmatrix}$.

The columns of this matrix are the coordinates of $f[3\pi/2](\mathbf{a})$ and $f[3\pi/2](\mathbf{b})$, respectively. Hence $f[3\pi/2](\mathbf{a}) = \frac{1}{2}\mathbf{a} - \frac{1}{2}\mathbf{b}$ and $f[3\pi/2](\mathbf{a}) = \frac{1}{2}\mathbf{a} + \frac{1}{2}\mathbf{b}$. From these we obtain

$$\begin{aligned} \mathbf{c} &= f[3\pi/2](\mathbf{d}) \\ &= n(\tfrac{1}{2}\mathbf{a} - \tfrac{1}{2}\mathbf{b}) + m(\tfrac{1}{2}\mathbf{a} + \tfrac{1}{2}\mathbf{b}) \\ &= \tfrac{1}{2}(n + m)\mathbf{a} + \tfrac{1}{2}(-n + m)\mathbf{b}. \end{aligned}$$

Solution 3.6

$\mathbf{I} - \mathbf{A}$ is the matrix

$$\begin{bmatrix} 1 - \frac{1}{2} & \frac{\sqrt{3}}{2} \\ -\frac{\sqrt{3}}{2} & 1 - \frac{1}{2} \end{bmatrix} = \begin{bmatrix} \frac{1}{2} & \frac{\sqrt{3}}{2} \\ -\frac{\sqrt{3}}{2} & \frac{1}{2} \end{bmatrix},$$

and it is easily verified that

$$\begin{bmatrix} \frac{1}{2} & -\frac{\sqrt{3}}{2} \\ \frac{\sqrt{3}}{2} & \frac{1}{2} \end{bmatrix} \begin{bmatrix} \frac{1}{2} & \frac{\sqrt{3}}{2} \\ -\frac{\sqrt{3}}{2} & \frac{1}{2} \end{bmatrix} = \begin{bmatrix} 1 & 0 \\ 0 & 1 \end{bmatrix},$$

hence $(\mathbf{I} - \mathbf{A})^{-1} = \mathbf{A}$. Thus, since $\mathbf{d} = (\mathbf{I} - \mathbf{A})\mathbf{c}$, it follows that

$$\begin{aligned} \mathbf{c} &= (\mathbf{I} - \mathbf{A})^{-1}\mathbf{d} \\ &= \mathbf{A}\mathbf{d} \\ &= r[\pi/3](\mathbf{d}). \end{aligned}$$

Solution 3.7

The 6-centres that lie on the basic parallelogram are $\mathbf{0}, \mathbf{a}, \mathbf{b}$ and $\mathbf{a} + \mathbf{b}$, and these are 3-centres as well. In addition, the points $(\mathbf{a} + \mathbf{b})/3$ and $2(\mathbf{a} + \mathbf{b})/3$ lie inside the basic parallelogram, and are 3-centres. With respect to $\{\mathbf{a}, \mathbf{b}\}$, their coordinates are $\left(\frac{1}{3}, \frac{1}{3}\right)$ and $\left(\frac{2}{3}, \frac{2}{3}\right)$, respectively. The matrix for $\mathbf{I} - \mathbf{A}$ is

$$\begin{bmatrix} 2 & 1 \\ -1 & 1 \end{bmatrix},$$

so that $r[(\mathbf{a}+\mathbf{b})/3, 2\pi/3] = t[\mathbf{d}]\, r[2\pi/3]$ where the coordinates of $\mathbf{d}$ in this basis are

$$\begin{bmatrix} 2 & 1 \\ -1 & 1 \end{bmatrix} \begin{bmatrix} \frac{1}{3} \\ \frac{1}{3} \end{bmatrix} = \begin{bmatrix} 1 \\ 0 \end{bmatrix}.$$

That is to say,

$$r\left[(\mathbf{a}+\mathbf{b})/3, 2\pi/3\right] = t[\mathbf{a}]\, r[2\pi/3].$$

Similarly,

$$r\left[2(\mathbf{a}+\mathbf{b})/3, 2\pi/3\right] = t[2\mathbf{a}]\, r[2\pi/3].$$

Now $r[4\pi/3] = (r[2\pi/3])^{-1}$ and so has the matrix

$$\mathbf{A}^{-1} = \begin{bmatrix} 0 & 1 \\ -1 & -1 \end{bmatrix}$$

with respect to $\{\mathbf{a}, \mathbf{b}\}$. So the matrix $\mathbf{I} - \mathbf{A}^{-1}$ is

$$\mathbf{I} - \mathbf{A}^{-1} = \begin{bmatrix} 1 & -1 \\ 1 & 2 \end{bmatrix},$$

and (arguing as above)

$$r[(\mathbf{a}+\mathbf{b})/3, 4\pi/3] = t[\mathbf{b}]\, r[4\pi/3],$$
$$r[2(\mathbf{a}+\mathbf{b})/3, 4\pi/3] = t[2\mathbf{b}]\, r[4\pi/3].$$

Solution 4.1

The obvious isomorphism (though not the only one) is

$$t_a^n t_b^m \mapsto (n, m) \quad (n, m \in \mathbb{Z}).$$

Solution 4.2

We may write $r[\mathbf{a}, \pi]$ as $t[2\mathbf{a}]\, r[\pi] = t_a^2 r$, and $r[\mathbf{b}, \pi]$ as $t[2\mathbf{b}]\, r[\pi] = t_b^2 r$.

Then

$$\begin{aligned} f_1 f_2 f_3 &= t_a^2 r t_b^2 t_b^2 r \\ &= t_a^2 r t_b^4 r \\ &= t_a^2 t_b^{-4} r^2 \\ &= t_a^2 t_b^{-4}, \end{aligned}$$

and this is xy where $x = t_a^2 t_b^{-4}$ and $y = e$.

Solution 4.3

If n and m are both even, we obtain the vertices of $\mathcal{T}$.

If n is odd and m is even, we obtain the midpoints of the long edges.

If n is even and m is odd, we obtain the midpoints of the short edges.

If n and m are both odd, we obtain the centres of the parallelograms.

Solution 4.4

Writing $r[\mathbf{a} + \mathbf{b}, \pi]$ as $t[2\mathbf{a} + 2\mathbf{b}]\, r$, and $q[\frac{1}{2}\mathbf{b}, 0]$ as $t[\mathbf{b}]\, q$, we obtain:

$$\begin{aligned} f_1 f_2 f_3 &= t[2\mathbf{a} + 2\mathbf{b}]\, r\, t[\mathbf{b}]\, q\, t[\mathbf{a}] \\ &= t[2\mathbf{a} + 2\mathbf{b}]\, t[-\mathbf{b}]\, r\, t[\mathbf{a}]\, q \\ &= t[2\mathbf{a} + 2\mathbf{b}]\, t[-\mathbf{b}]\, t[-\mathbf{a}]\, rq \\ &= t[\mathbf{a} + \mathbf{b}]\, rq \\ &= xy, \quad \text{where} \quad x = t[\mathbf{a} + \mathbf{b}] = t_a t_b \quad \text{and} \quad y = rq. \end{aligned}$$

Solution 4.5

There are four orbits of 2-centres under the action of $\Gamma(L)$. As with the parallelogram lattice, they have representatives whose position vectors are $\mathbf{0}, \frac{1}{2}\mathbf{a}, \frac{1}{2}\mathbf{b}$ and $\frac{1}{2}\mathbf{a} + \frac{1}{2}\mathbf{b}$.

Perhaps the easiest way to see this is to recognize that the direct symmetry group, $\Gamma^+(L)$, is simply the symmetry group of the parallelogram lattice, which partitions the 2-centres into the above four orbits; and also to recognize that the indirect symmetries in $\Gamma(L)$ never send centres in one orbit to centres in another orbit.

Solution 4.6

The symmetries that fix $\mathbf{c}$ are $e, t_a^n t_b^m r, t_b^m q$ and $t_a^n rq$. Alternatively, they could be written as $e, t[n\mathbf{a} + m\mathbf{b}]\, r, t[m\mathbf{b}]\, q$ and $t[n\mathbf{a}]\, rq$.

Solution 4.7

The line l' is the axis of $q' = rq$. Now $\text{Orb}(l') = \{g(l') : g \in \Gamma(L)\}$, and $g(l')$ is the axis of the reflection $grqg^{-1}$. From the relations, you can check that $grqg^{-1} = rq$ when g is t_b, r or q. So we need only look at the conjugates when $g = t_a^n$. Here we obtain

$$t_a^{2n} rq,$$

and this is the reflection in the line parallel to l' which passes through the lattice point $n\mathbf{a}$. Thus $\text{Orb}(l')$ consists of all the lines parallel to l' which pass through the lattice points $n\mathbf{a}$ ($n \in \mathbb{Z}$).

Solution 4.8

The line l' is fixed by the elements e, r, q, t_b and all combinations of these. Thus, letting $T_b = \{t_b^m : m \in \mathbb{Z}\}$, the stabilizer of l' is

$$\text{Stab}(l') = \{xy : x \in T_b,\ y \in D_2;\ rt_b = t_b^{-1} r,\ qt_b = t_b^{-1} q\}.$$

Solution 4.9

$\text{Stab}(\mathbf{0}) = D_2 = \{e, r, q, rq\}$.

$\text{Stab}\left(\frac{1}{2}\mathbf{a}\right) = \{e, t_a r\}$, isomorphic to C_2.

$\text{Stab}\left(\frac{1}{2}\mathbf{a} + \frac{1}{2}\mathbf{b}\right) = \{e, t_a t_b r, q, t_a t_b rq\}$, isomorphic to D_2.

Solution 4.10

If $\mathbf{x}$ is the centre of a rotation r' of order 4 and g an element of $\Gamma(L)$, then $g(\mathbf{x})$ is the centre of the rotation $gr'g^{-1}$, which is also of order 4. Hence $\Gamma(L)$ acts on the 4-centres.

The lattice points will belong to one orbit and the points $\{(n + \frac{1}{2})\mathbf{a} + (m + \frac{1}{2})\mathbf{b} : n, m \in \mathbb{Z}\}$, which are the centres of the squares (i.e. the basic parallelograms), will belong to one orbit. There is no symmetry in $\Gamma(L)$ which maps $\mathbf{0}$ to $\frac{1}{2}\mathbf{a} + \frac{1}{2}\mathbf{b}$, so these orbits are distinct. Thus there are exactly two orbits of 4-centres.

$\text{Stab}(\mathbf{0}) = D_4 = \{e, r, r^2, r^3, q, rq, r^2 q, r^3 q\}$.

$\text{Stab}(\frac{1}{2}\mathbf{a} + \frac{1}{2}\mathbf{b}) = \{e, t_a r, t_a t_b r^2, t_b r^3, t_b q, rq, t_a r^2 q, t_a t_b r^3 q\}$, also isomorphic to D_4.

Solution 4.11

The 2-centres which are not 4-centres are the points of the form
$\frac{1}{2}\mathbf{a} + n\mathbf{a} + m\mathbf{b}$ and $\frac{1}{2}\mathbf{b} + n\mathbf{a} + m\mathbf{b}$ (that is, $(n+\frac{1}{2})\mathbf{a} + m\mathbf{b}$ and
$n\mathbf{a} + (m+\frac{1}{2})\mathbf{b}$). These form a single orbit, since $r(\frac{1}{2}\mathbf{a}) = \frac{1}{2}\mathbf{b}$.

$\text{Stab}(\frac{1}{2}\mathbf{a}) = \{e, t_a r^2, q, t_a r^2 q\}$, isomorphic to D_2.

Solution 4.12

There is a single orbit of 6-centres, consisting of the lattice points.

$\text{Stab}(\mathbf{0}) = D_6 = \{e, r, r^2, r^3, r^4, r^5, q, rq, r^2 q, r^3 q, r^4 q, r^5 q\}$.

There is a single orbit of 3-centres which are not 6-centres, consisting of the points of the form $(n + \frac{1}{3})\mathbf{a} + (m + \frac{1}{3})\mathbf{b}$ and $(n + \frac{2}{3})\mathbf{a} + (m + \frac{2}{3})\mathbf{b}$. Let
$\mathbf{c} = \frac{1}{3}(\mathbf{a} + \mathbf{b})$. Then $\mathbf{c}$ is fixed by the symmetries $e, t_a r^2, t_b r^4, rq, t_a r^3 q$ and
$t_b r^5 q$, and these form the stabilizer $\text{Stab}(\mathbf{c})$, which is isomorphic to D_3.

The 2-centres which are not 6-centres consist of the points of the form
$(n + \frac{1}{2})\mathbf{a} + m\mathbf{b}$, $n\mathbf{a} + (m + \frac{1}{2})\mathbf{b}$ and $(n + \frac{1}{2})\mathbf{a} + (m + \frac{1}{2})\mathbf{b}$. They form a single
orbit, since $r(\mathbf{a}/2) = \mathbf{b}/2$ and $t_a r(\mathbf{b}/2) = (\mathbf{b} - \mathbf{a})/2 + \mathbf{a} = (\mathbf{a} + \mathbf{b})/2$. The
stabilizer of the point $\mathbf{a}/2$ is $\{e, t_a r^3, q, t_a r^3 q\}$, which is isomorphic to D_2.

Solution 5.1

As $\|\mathbf{b}\| > 0$, we may divide the inequalities in Condition (b) by $\|\mathbf{b}\|$:

$$-\frac{\|\mathbf{a}\|}{2\|\mathbf{b}\|} \leq \cos\theta \leq \frac{\|\mathbf{a}\|}{2\|\mathbf{b}\|}.$$

But by Condition (a), $\|\mathbf{a}\|/\|\mathbf{b}\| \leq 1$, and so

$$-\tfrac{1}{2} \leq \cos\theta \leq \tfrac{1}{2}.$$

Thus, since θ is chosen to lie between 0 and π, we have

$$\pi/3 \leq \theta \leq 2\pi/3.$$

> This is essentially a rewording of the proof of part (c) of the theorem on page 17.

Solution 5.2

If $\mathbf{b}$ lies in the 'lower' half of the region, replace it by $-\mathbf{b}$; clearly we still
have a reduced basis.

Solution 5.3

Since $\{\mathbf{a}, \mathbf{b}\}$ is a reduced basis, $\mathbf{b}$ lies in the minimality region, and so
$-\frac{1}{2} \leq x \leq \frac{1}{2}$. Thus, $-1 \leq 2x \leq 1$. But if $2x$ were to take any value in
this interval other than $-1, 0$ or 1, then $\mathbf{c}$ would be a non-zero vector of smaller
magnitude than $\mathbf{a}$, contradicting our choice of $\mathbf{a}$ to be of least magnitude in L.

Solution 5.4

We are now dealing with the case $\mathbf{c} = \mathbf{b} + q(\mathbf{b}) = -\mathbf{a}$, and so $q(\mathbf{b}) = -\mathbf{a} - \mathbf{b}$.

The transition matrix from $\{-\mathbf{b}, \mathbf{a} + \mathbf{b}\}$ to $\{\mathbf{a}, \mathbf{b}\}$ has integer entries and
determinant 1, so $\{-\mathbf{b}, \mathbf{a} + \mathbf{b}\}$ is a basis for L. Moreover,

$$q(-\mathbf{b}) = -(-\mathbf{a} - \mathbf{b}) = \mathbf{a} + \mathbf{b},$$
$$\begin{aligned}q(\mathbf{a} + \mathbf{b}) &= \mathbf{a} + (-\mathbf{a} - \mathbf{b}) \\ &= -\mathbf{b},\end{aligned}$$

and so f is rhombic.

Solution 5.5

As before, the fact that **b** lies in the minimality region and **a** is of least magnitude shows that $\mathbf{d} = \mathbf{0}$ or $\pm \mathbf{a}$.

Subcase (a) $\mathbf{d} = \mathbf{0}$.

Since **d** was defined to be $\mathbf{b} - q(\mathbf{b})$, we have
$$q(\mathbf{b}) = \mathbf{b}.$$

As we are dealing with the case in which $q(\mathbf{a}) = -\mathbf{a}$, we exchange basis elements; the basis $\{\mathbf{b}, \mathbf{a}\}$ then tells us that f is rectangular.

Subcase (b) $\mathbf{d} = \mathbf{a}$.

In this case,
$$\mathbf{b} - q(\mathbf{b}) = \mathbf{a},$$
and so
$$q(\mathbf{b}) = \mathbf{b} - \mathbf{a}.$$
Then
$$q(\mathbf{b} - \mathbf{a}) = (\mathbf{b} - \mathbf{a}) - (-\mathbf{a})$$
$$= \mathbf{b},$$
and so, using the basis $\{\mathbf{b}, \mathbf{b} - \mathbf{a}\}$, we see that f is rhombic.

Subcase (c) $\mathbf{d} = -\mathbf{a}$.

In this case,
$$\mathbf{b} - q(\mathbf{b}) = -\mathbf{a},$$
and so
$$q(\mathbf{b}) = \mathbf{b} + \mathbf{a}.$$
Then
$$q(\mathbf{b} + \mathbf{a}) = (\mathbf{b} + \mathbf{a}) + (-\mathbf{a})$$
$$= \mathbf{b},$$
and so, using the basis $\{\mathbf{b}, \mathbf{b} + \mathbf{a}\}$, we once again see that f is rhombic.

OBJECTIVES

After you have studied this unit, you should be able to:

(a) decide whether a pair of vectors is a basis for a given lattice and be able to find a reduced basis;

(b) write a given symmetry of a lattice in standard form and calculate the composite of any two of its symmetries;

(c) find the centre of any rotation of a lattice given in standard form, and find the reflection and glide axes;

(d) find the type of a lattice with a given basis and specify its symmetries;

(e) write down each lattice group in terms of its generators and relations as given in the text;

(f) find the orbits and stabilizers of 2-centres, 3-centres, 4-centres and 6-centres;

(g) find the orbits and stabilizers of reflection and glide axes;

(h) distinguish between those indirect symmetries of a lattice that are rectangular and those that are rhombic;

(i) understand why there are only five types of plane lattice;

(j) relate the type of a lattice to the position of **b** within the minimality region, where $\{\mathbf{a}, \mathbf{b}\}$ is a reduced basis.

INDEX

2-centres 27
3-centres 27
4-centres 27
6-centres 27
axes
    glide 31
    reflection 31
basic parallelogram 12
basis 8
centred rectangular lattice 31
crystallographic restriction 25
essential glide reflection 20
glide axes 31
glide reflection of the lattice 19
hexagonal lattice 33
identity of lattices 11

indirect symmetry
    rectangular 46
    rhombic 46
inessential glide reflection 20
integer combination 6
lattice 6
lattice of integer points 8
minimality conditions 15
minimality region 45
orthogonal basis 28
parallelogram lattice 27
plane lattice 6
rectangular lattice 28
reduced basis 17
reflection axes 31
reflection component 19

reflection of the lattice 19
rhombic lattice 30
rotation of the lattice 18
square lattice 31
strict triangle inequality 16
transition matrix 11
translation component 19
translation of the lattice 18
translation
    magnitude of 18
    parallel 18
    proper 18
triangle inequality 16
two-dimensional lattice 6